Current State of Art in Artificial Intelligence and Ubiquitous Cities

Rita Yi Man Li · Kwong Wing Chau ·
Daniel Chi Wing Ho
Editors

Current State of Art in Artificial Intelligence and Ubiquitous Cities

Springer

Editors
Rita Yi Man Li
Department of Economics and Finance
Sustainable Real Estate Research Center
Hong Kong Shue Yan University
Hong Kong, Hong Kong

Kwong Wing Chau
Department of Real Estate and Construction
The University of Hong Kong
Hong Kong, Hong Kong

Daniel Chi Wing Ho
Department of Environment, Technological
and Higher Education
Institute of Hong Kong
Hong Kong, Hong Kong

ISBN 978-981-19-0736-4 ISBN 978-981-19-0737-1 (eBook)
https://doi.org/10.1007/978-981-19-0737-1

This Springer imprint is published by the registered company Springer Nature Singapore Pte Ltd.
The registered company address is: 152 Beach Road, #21-01/04 Gateway East, Singapore 189721, Singapore

Contents

Chapter 1
Artificial Intelligence for Heritage Conservation: A Case Study of Automatic Visual Inspection System

Lukman E. Mansuri and D. A. Patel

Abstract Heritage plays a fundamental role in maintaining the identity of the city. The character and historical significance of any city is portrayed through its built and cultural heritage. Heritage is the essence of society, and it should be conserved and preserved for future generations. The latest technologies in this digital era become a key governing factor for the conservation and preservation of heritage. Artificial intelligence (AI) based automation is the emerging technology for the conservation and preservation of heritage. The application of AI in heritage conservation is divided into four clusters, (i) AI for visual inspection and structural health monitoring, (ii) AI for digital modeling, (iii) AI for intangible heritage, and (iv) AI for cultural heritage. This chapter summarizes various techniques and algorithms of AI used for a variety of applications in heritage conservation. The chapter also outlines some of the notable developments of AI for heritage conservation and presents the case study of the development of automatic visual inspection system for heritage structures. Lastly, the chapter concludes with the framework of AI for heritage conservation in ubiquitous cities and the way forward.

1 Introduction

Heritage plays a fundamental role in reinforcing the identity of the city. The cultural of any city is reflected by its historical significance. Hence, heritage should be conserved and preserved for future generations. Heritage buildings and monuments have its own significance in history, culture, politics, religion, and economics. Conservation of heritage is the part of United Nations (UN) sustainable development goals (SDGs) for development agenda 2030 (Petti et al., 2020). The latest technologies in this digital era become a key governing factor for the conservation and preservation of heritage. Preservation of heritage leads to the social and economic growth of the city by attracting and promoting the tourism (Binns & Nel, 2002). Hence, heritage conservation is the latent part of the development of the ubiquitous city. In the current

L. E. Mansuri (✉) · D. A. Patel
Department of Civil Engineering, Sardar Vallabhbhai National Institute of Technology, Surat, India

R. T. M. Li et al. (eds.), *Current State of Art in Artificial Intelligence and Ubiquitous Cities*, https://doi.org/10.1007/978-981-19-0737-1_1

decade of the industry 4.0 revolution, digital technologies lead towards the era of self-intelligence for the development of cities (Newman et al., 2020). The ubiquitous city is a sustainable and smart city with an established and effective urban management system using information and communication technologies (ICTs). ICT is applicable in various elements of urban management such as planning, governance, health, welfare, education, and heritage conservation (Lee et al., 2008).

Preservation and conservation of heritage is a global challenge (Udeaja et al., 2020). Heritage buildings and monuments differ from modern buildings in terms of their construction style, design, structural systems, and building materials. The functional performance of heritage buildings decreases due to degradation and decay of material, weather effects, environmental degradation, earthquake, and vandalism by visitors and tourists (Taboroff, 2000). These effects result in various defects in the structure, such as crack, spalling, effloresce, stains, abrasion, discoloration, and fungal growth (Directorate General: Conservation of Heritage Buildings, 2013). Special attention and care are required to retain the physical and architectural character of heritage buildings. Without a proper understanding of the present condition of the heritage structure, conservation becomes a challenging task. Condition assessment is carried out to examine the structure's present condition in terms of knowing various types and severity of defects in the structure. Visual inspection is the primary stage of condition assessment of heritage structure followed by detailed structural health monitoring and diagnosis. Repair and maintenance strategies are decided based on the results of condition assessment and structural health monitoring (Riggio et al., 2018).

Traditionally, the condition assessment and structural health monitoring are conducted through manual and equipment based methods. Manual methods require highly skilled staff and time to assess the structure's condition, which results in high staff cost and prone to human errors (Mansuri & Patel, 2020a). Also, the manual method becomes impractical at some places due to the heritage buildings' structural and architectural complexity. The condition assessment and structural health monitoring (SHM) are now supplemented by various advanced equipment and sensors (Aras et al., 2011; Boscato et al., 2016; Flores-Colen et al., 2009; Gattulli & Chiaramonte, 2005; Gentile & Gallino, 2008; Hallermann et al., 2015; Ibarra-Castanedo et al., 2017; O'Byrne et al., 2013; Pérez-Gracia et al., 2013; Pieraccini et al., 2005; Reyno et al., 2018). These equipment and sensors are very accurate in data collection and advantageous for monitoring of the structures. Despite various advantages, the equipment and sensors have multiple limitations, such as huge installation and operation costs, and longtime data collection (Barreira & Freitas, 2007; Flores-Colen et al., 2011; Ibarra-Castanedo et al., 2017). After data collection by sensors and equipment, the post-processing also requires highly skilled experts and advanced software systems that make it more costly (Huang et al., 2014). Due to such limitations of traditional condition survey system, heritage structures' conservation and preservation is delayed and become expensive and time-consuming. Therefore, there is need of advanced technology that overcomes limitations of the conventional methods of heritage conservation.

Nowadays, the application of ICT and automation of processes is adopted and promoted by governments in cities. Artificial intelligence (AI) based automation is the emerging technology for the conservation and preservation of heritage (Mansuri et al., 2019). AI can overcome the limitations of conventional practices in the field of heritage conservation. Various areas such as medical diagnosis (Mohanty et al., 2016; Ramcharan et al., 2017), infrastructure condition assessment (Gopalakrishnan et al., 2017; Kumar et al., 2020), and information science (Saadi & Belhadef, 2020) have achieved remarkable benchmarks by using AI. AI has also been applied in the fields of cultural and built heritage conservation and preservation. This chapter outlines some of the notable developments of AI for heritage conservation and presents the case study of the development of automatic visual inspection system for heritage structures.

This chapter contains five sections. Section one represents the introduction of this chapter. Section two outlines various applications and techniques of AI for heritage conservation. The third section presents the case study of the development of an automatic visual inspection system. The framework for AI-based heritage conservation in ubiquitous cities is presented in section four. Lastly, section five gives the concluding remarks of this chapter.

2 AI for Heritage Conservation

In recent times, artificial intelligence has shown promising results in the application of heritage conservation and preservation. The application of AI in heritage conservation can be divided into four clusters as shown in Fig. 1, (i) AI for visual inspection and structural health monitoring, (ii) AI for digital modeling, (iii) AI for intangible heritage, and (iv) AI for cultural heritage. AI has been used for heritage conservation directly and indirectly. For example, automatic defect detection is the direct area of

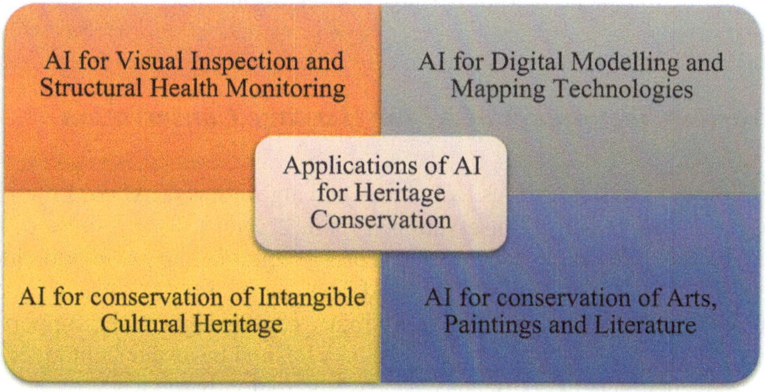

Fig. 1 Application of AI for heritage conservation (*source* authors' figure)

conservation while, quantification of damage is the indirect application of AI for heritage conservation.

2.1 Various Application Areas of AI for Heritage Conservation

AI has proven the capabilities for inspection and SHM of various civil infrastructures such as roads, bridges, tunnels, railway, and buildings (Koch et al., 2015; Ouma & Hahn, 2017; Yeum & Dyke, 2015). The application of AI for heritage is the recent trend of research. Various AI methods have been applied to identify and detect damage of efflorescence, spalling, lost material, and cracks are the current focus areas for automatic condition assessment in heritage monuments (Valero et al., 2019). The development of AI-based automated inspection systems is also a growing area of heritage conservation research (Mansuri & Patel, 2020b). Automatic and semi-automatic modeling using deep learning and other AI techniques is the second area of application (Macher et al., 2014). Automatic generation of models from point cloud data is an emerging area of research for heritage building information modelling (HBIM) based conservation. Automatic generation of HBIM leads to faster and cost-effective conservation practice. Researches built various applications and algorithms to preserve intangible cultural heritage such as classical dance, yoga, language, and music (Berman & James, 2015; Chiddarwar et al., 2020; Saadi & Belhadef, 2020). These AI-based algorithms and applications help manage, learn, innovate, and disseminate the intangible cultural heritage. Conservation arts, artifacts, paintings, and images are the fourth central area of AI for heritage conservation. Digital preservation of arts and images using image in painting enhances the visual experience (Belhi et al., 2020). Detection of cracks in the old precious paintings and artifacts is also the advanced application of AI for the conservation and preservation of heritage. Hence, AI has enormous potential for further development and application in a niche field of heritage conservation.

2.2 Various Techniques of AI for Heritage Conservation

This section summarizes various techniques and algorithms of AI used for a variety of applications in heritage conservation. Figure 2 shows the multiple techniques used in the field of heritage conservation. The AI techniques are mainly divided into two categories, machine learning (ML) and deep learning (DL). ML consists of use of artificial neural network (ANN), fuzzy logic, genetic algorithms, support vector machine (SVM), support vector regression (SVR), and bayesian techniques. While convolutional neural network (CNN), deep CNN, region-based CNN are considered as the deep learning techniques.

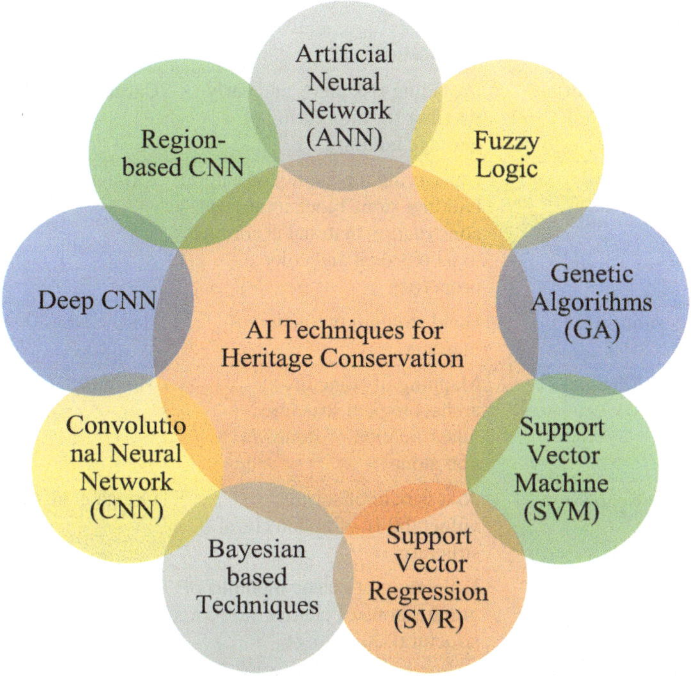

Fig. 2 Various AI techniques for heritage conservation (*source* authors' figure)

A list of various techniques along with their specific applications is listed in Table 1. These techniques are applied to heritage for direct and indirect applications. Several techniques are used for visual inspection and condition survey, while some techniques are being used for detailed structural health monitoring of heritage monuments.

In line of above stated applications of AI for heritage, the following section presents the case study of development of automatic visual inspection system for heritage structures using deep learning.

3 Development of Automatic Inspection System

This section explains the formulation of an automatic visual inspection system by developing automatic defect detection model. The automatic defect detection model can be developed using advanced object detection algorithm of deep learning. Based on the literature review and the accuracy of the model, Faster R-CNN (faster region-based convolutional networks) (Ren et al., 2015) model of deep learning is selected to develop automatic defect detection system.

The automatic visual inspection system is developed for the British cemetery site (Fig. 3) located in the city of Surat, India. The city of Surat is having reach built

Table 1 Applications of AI techniques

AI technique	Application	References
Artificial Neural Network (ANN)	Providing risk alerts and early warning for monitoring of heritage building	Carnimeo et al. (2015)
Fuzzy Logic	Classifying the quality of natural building stone block as ancient construction material according to its physical and color properties	Akkoyun & Toprak (2012)
Genetic Algorithms (GA)	Bond strength prediction of masonry	Harp et al. (2009)
Support Vector Machine (SVM)	Mapping of stone in archaeological structure (classification of stone v/s non-stone)	Monna et al. (2020)
Support Vector Regression (SVR)	Assessment of seismic vulnerability of architectural structure	Ortega et al. (2019)
Bayesian based Techniques	Estimation of elastic modulus of granite stone of church for structural assessment	Ramos et al. (2015)
Convolutional Neural Network (CNN)	Location estimation of missing component in heritage structure	Zou et al. (2019)
Deep CNN	Detection of defective bricks in the historic masonry wall of museum	Wang et al. (2019)
Region-based CNN	Development of automatic visual inspection system for built heritage	Mansuri and Patel (2020a)

and cultural heritage. The city of Surat was an important trading town through sea trade in the sixteenth, seventeenth and eighteenth centuries. Surat became the most important trade link between India and many other countries and was at the height of prosperity till the rise of Bombay port in the seventeenth and eighteenth centuries (Parveen, 2014). Hence, Surat city is having four nationally protected heritage sites built by the traders from England, Netherlands, and Armenia during seventeenth and eighteenth century. There is a present need to conserve these national heritage sites. Therefore, this study attempted to develop the automatic inspection system for heritage structures.

Fig. 3 Surat city and site of British cemetery (*source* authors' figure)

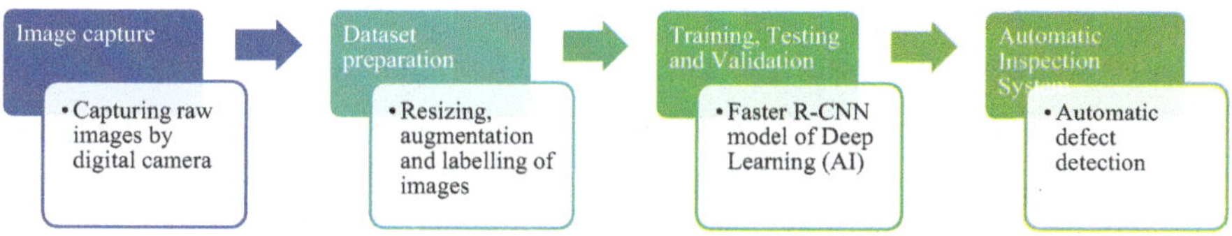

Fig. 4 Research method for development of automatic inspection system (*source* authors' figure)

3.1 Research Method

The research method for development of automatic inspection system is illustrated in Fig. 4. The raw images were collected using digital camera from the British cemetery (national heritage site in the city of Surat, India). The raw images then resized, augmented and labeled to create training, validation and testing dataset. Then the dataset has been trained using Faster R-CNN on Google Colab cloud GPU. Automated defect detection system is developed using trained classifier.

3.2 About Deep Learning Model: Faster R-CNN

Faster R-CNN is widely used model for object detection. It proposes a way to speed up the calculation of region proposals by establishing a Region Proposal Network (RPN). The RPN is a fully connected convolutional network that trains high-quality region proposals by an end-to-end manner. Trained convolution features by RPN

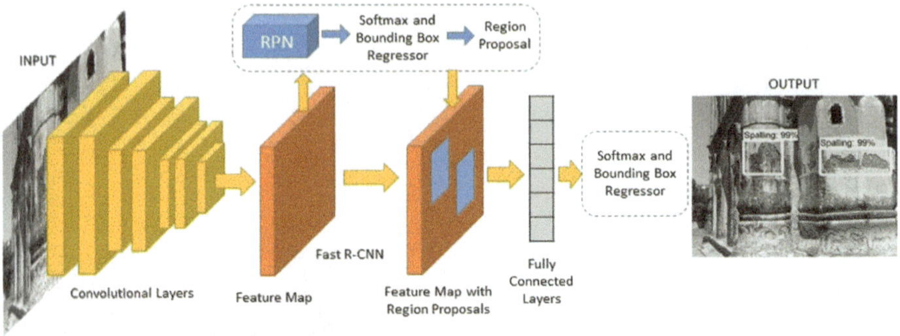

Fig. 5 Conceptual architecture of Faster R-CNN (*source* authors' figure)

are then shared with Fast R-CNN. The conceptual architecture of Faster R-CNN is illustrated in Fig. 5. Faster R-CNN mainly composed by two stages, (i) RPN boxes extraction, and (ii) Fast R-CNN detection (Ren et al., 2015).

Bounding boxes is extracted by using RPN which is a fully convolutional neural network, then, object proposal is detected and identified by Fast R-CNN based on extracted proposal by RPN.

The Faster R-CNN is principally divided into four major components as follows;

(1) Convolution layers: Faster R-CNN is using a set of basic convolutional layers, ReLU (rectified linear unit) function for activation and pooling layers for extraction of feature maps in image. Feature map are shared with subsequent fully connected and RPN layers (Nair & Hinton, 2010).
(2) Region proposal networks (RPN): generation of region proposals are done by region proposal networks. SoftMax regression is used by this layer to determine the foreground or background anchors and uses bounding box regression by correcting the anchors to get accurate proposals (Ren et al., 2015).
(3) Region of Interest (RoI) pooling: Input feature maps and proposals are collected by this layer. It extracts the feature maps after synthesizing the information and sending it to related fully connected layer for determining the object categories (Mansuri & Patel, 2020b).
(4) Classification: The classification of proposals is calculated using proposal feature maps in Faster R-CNN and parallel it uses the bounding box regression to obtain coordinates of detection boxes precisely.

3.3 Dataset Preparation

For image based deep learning, good image dataset is the preliminary need. The images for this study were taken from the British Cemetery located in the historic city of Surat, India, as shown in Fig. 3. This site is protected as the national heritage site under Archeological Survey of India (ASI). This site consists of eighteen monumental structures in the form of tombs. These monuments were constructed by British rulers of India during 1649–1811 AD (Scarre & Roberts, 2005). All the monuments were constructed using brick masonry and covered with yellow stucco. Spalling is found as the major and most common surface defect. The spalling is then divided into two categories: 'spalling' and 'exposed bricks'. Thus, this study is focusing on automatic detection of these two surface defects.

Total 300 images of 3120 × 4160 pixels were taken by a mobile digital camera. The images were taken to create the variety of dataset in terms of varying backgrounds, lighting conditions, and different distances from camera. Variety of backgrounds in the image dataset is required for precise detection of defects in various conditions. These 300 raw images were then resized to smaller size of 800 × 600 pixels for reducing the training time. To develop main database, augmentation was applied to 300 images and total 600 images of 800 × 600 pixels were created by horizontal flipping and brightness changing. 80% of images kept for training and 10% for validation and 10% for testing. Thus, from the dataset of 600 images, 480 images kept for training, 60 for validation and 60 for testing.

For training of defect detection system, the defect present in image has been manually identified first. The manual identification known as labelling which serve as ground truth images that are used to learn defects from image. The images are labelled for defect types in two categories, spalling and exposed bricks. is the images are labeled by open source labelling software tool LabelImg (Tzutalin: LabelImg, 2018). The labelling tool allows easy labelling of images by interactive interface. The output of labelling tool is into extensible markup language (XML) file corresponding to labelled image in the pascal visual object classes (PascalVOC) format. The XML file of each image contains information about image such as name, size and bounding box coordinates.

3.4 *Training, Testing and Validation*

This section explains how to train, test and validate the dataset for the development of model. The experiment were conducted using open-source Faster R-CNN model on Google Colab cloud GPU. Object detection training consists of three steps: (i) performing forward pass, (ii) calculating loss, and (iii) weight updating of network (Ren et al., 2015). In forward pass, training image which is annoted by ground truth that is passed through the model and output results are calculated. These outputs are compared with ground truth to calculate loss. The main goal of training is to minimize the loss. The training step is to be continued until the loss is minimum and or almost constant. Training steps are the important hyper-parameter which directly affects training time and precision of detection. With less number of iterations, the training time is short but the defect detection precision is very low. Training loss can be minimized and stabled with more number of training steps.

In this study, the optimum number of training steps has been found by trial-and-error method with the training steps 5000–45,000. The minimum loss was recorded on 38,500 steps. Thus, the final training of whole dataset is achieved on 38,500 steps which have taken 6 h of training time. To determine the accuracy of detection, intersection over union (IoU) matrix has been used for this study. The IoU is the measure of degree of overlap between ground truth and detected bounding box. The threshold value of IoU is selected as 0.50 for experiment in this study. The testing is conducted on the test dataset to check the accuracy of the defect detection. 0.891mAP is recorded as the accuracy of defect detection.

The final dataset was trained and the classifier was used to detect defects from the real-life site images. All the results generated by the developed classifier are shown in Fig. 5. The confidence level of detecting multiple defects are ranging from 50 to 100%. The detection confidence level is shown on the top corner of each detected bounding box as shown Fig. 6. Various site conditions are tested for the detection of defects. In Fig. 6a spalling is detected at three places with a confidence level of 100%. Figure 6b illustrates the detection of both categories of defects; spalling and exposed bricks in the same image. The place of bounding box for each defect is corrected as per visual understanding. In some parts of the structure, efflorescence is present that has same colour as of spalling. In spite of having same colour of efflorescence and spalling, model is able to detect spalling. In static 2D images, the appearance of spalling and effloresces are very similar. This similarity may lead to misclassification and inaccurate detection. This automatic defect detection system can be used as primary visual inspection system for inspection of monuments of British cemetery.

4 AI-Based Heritage Conservation in Ubiquitous Cities: The Way Forward

In the upcoming future, heritage conservation in ubiquitous cities will have the rapid digital revolution. More advanced technologies based on AI will become part of heritage conservation, preservation, management, and promotion. As illustrated in Fig. 7, six focus areas have been identified as critical directions that can overcome the current manual conventional heritage conservation practices. The first three areas (automatic inspection and structural health monitoring, repair, maintenance and rehabilitation planning, and digitization of heritage) will support the conservation and preservation of heritage. Another three areas (virtual technology, digital twins, and 3D printing) will boost dissemination and heritage promotion.

As discussed in the earlier section, AI-based automatic inspection and SHM (structural health monitoring) can provide data to the intelligent maintenance management system. The smart maintenance management system can provide better alternative for the repair and rehabilitation of heritage. This process can also be supported by the digitalized model of the heritage structure in terms of HBIM (heritage building information modelling). Automation in modelling can also help in the faster generation of HBIM to facilitate the maintenance management process.

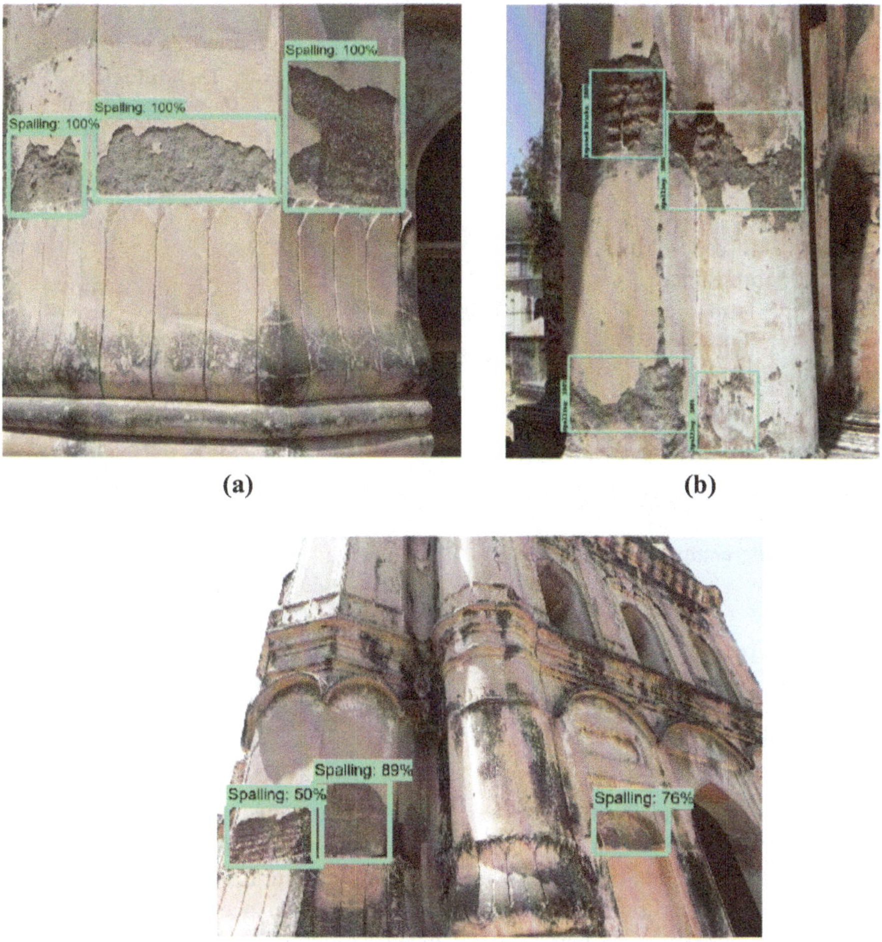

(c)

Fig. 6 Automatic defect detection results (*source* authors' photo)

After the preservation and conservation, AI-based systems can also support in dissemination and promotion of heritage. Virtual technologies such as AR (augmented reality), VR (virtual reality), and MR (mixed reality) for museums and libraries can be the primary tools for tourist attractions. The digital twins can be useful for various simulations and analysis of heritage structure to predict the risk of uncertain events and structural failure. 3D printing can support the replacement of missing elements of the structure and the creation of miniature models for promotion and tourist attractions.

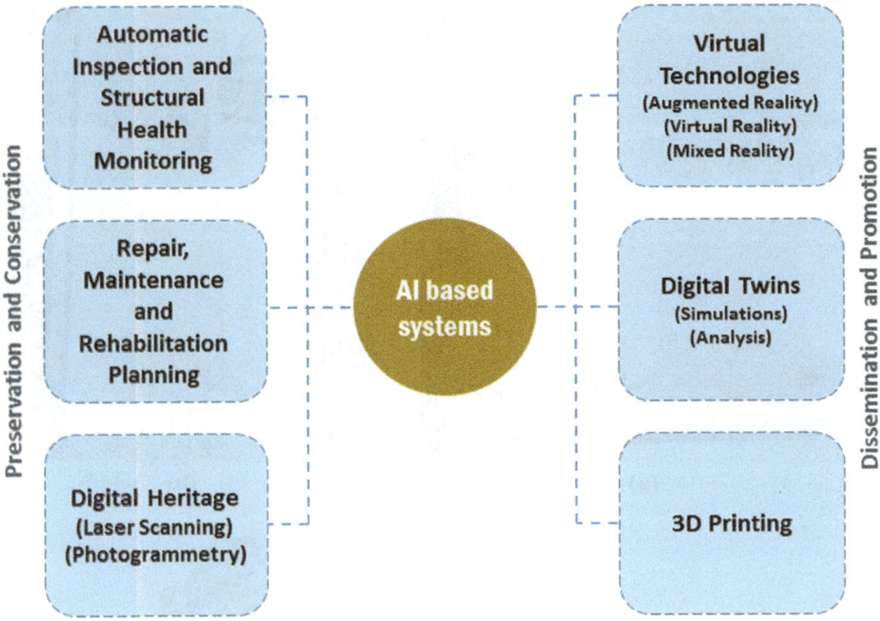

Fig. 7 AI for Heritage Conservation: Way forward (*source* authors' photo)

5 Conclusion

This chapter presents the brief review of various applications and techniques of artificial intelligence for heritage conservation. Various application and techniques of AI are advantageous for condition assessment, structural health monitoring and promotion of the heritage. The time and cost of maintenance and conservation of heritage can be reduced by using the advanced AI techniques for various purposes. The AI techniques enables to diagnose the heritage structure without contacting the vulnerable parts of the structures. Hence, the non-contact based remote diagnostics of heritage structure become efficient and accurate.

The case study in this chapter describes the specific use of AI for inspection and condition assessment of heritage monuments. The automatic defect detection system will be useful for the conservation authorities to inspect the monuments on the regular basis. The conservation authorities can plan their financial budgets for maintenance based on the automatic inspection system. Finally, six focus areas have been identified as critical directions that can overcome the current manual and conventional practice of heritage conservation. In short, this chapter explains the application of AI in heritage conservation in urban areas and presents the methodology to develop the automatic visual inspection system.

References

Akkoyun, O., & Toprak, Z. F. (2012). Fuzzy-based quality classification model for natural building stone blocks. *Engineering Geology, 133*, 66–75.

Aras, F., Krstevska, L., Altay, G., & Tashkov, L. (2011). Experimental and numerical modal analyses of a historical masonry palace. *Construction and Building Materials, 25*, 81–91.

Barreira, E., & de Freitas, V. P. (2007). Evaluation of building materials using infrared thermography. *Construction and Building Materials, 21*, 218–224.

Belhi, A., Bouras, A., Al-Ali, A. K., & Foufou, S. (2020). A machine learning framework for enhancing digital experiences in cultural heritage. *Journal of Enterprise Information Management.*

Berman, A., & James, V. (2015). Kinetic imaginations: exploring the possibilities of combining AI and dance. In *Twenty-Fourth International Joint Conference on Artificial Intelligence.*

Binns, T., & Nel, E. (2002). Tourism as a local development strategy in South Africa. *Geographical Journal, 168*, 235–247.

Boscato, G., Dal Cin, A., Ientile, S., & Russo, S. (2016). Optimized procedures and strategies for the dynamic monitoring of historical structures. *Journal of Civil Structural Health Monitoring, 6*, 265–289.

Carnimeo, L., Foti, D., & Vacca, V. (2015). On damage monitoring in historical buildings via neural networks. In *2015 IEEE Workshop on Environmental, Energy, and Structural Monitoring Systems (EESMS) Proceedings* (pp. 157–161). IEEE.

Chiddarwar, G. G., Ranjane, A., Chindhe, M., Deodhar, R., & Gangamwar, P. (2020). AI-based yoga pose estimation for android application.

Directorate General: Conservation of Heritage Buildings - A Guide. (2013).

Flores-Colen, I., de Brito, J., & de Freitas, V. (2009). Expected render performance assessment based on impact resistance in situ determination. *Construction and Building Materials, 23*, 2997–3004.

Flores-Colen, I., Manuel Caliço Lopes de Brito, J., & Peixoto de Freitas, V.: On-site performance assessment of rendering façades for predictive maintenance. *Structural Survey, 29*, 133–146 (2011).

Gattulli, V., & Chiaramonte, L. (2005). Condition assessment by visual inspection for a bridge management system. *Computer-Aided Civil and Infrastructure Engineering, 20*, 95–107.

Gentile, C., & Gallino, N. (2008). Condition assessment and dynamic system identification of a historic suspension footbridge. *Structural Control and Health Monitoring: The Official Journal of the International Association for Structural Control and Monitoring and of the European Association for the Control of Structures, 15*, 369–388.

Gopalakrishnan, K., Khaitan, S. K., Choudhary, A., & Agrawal, A. (2017). Deep Convolutional Neural Networks with transfer learning for computer vision-based data-driven pavement distress detection. *Construction and Building Materials, 157*, 322–330.

Hallermann, N., Morgenthal, G., & Rodehorst, V. (2015). Vision-based monitoring of heritage monuments: Unmanned Aerial Systems (UAS) for detailed inspection and high-accuracy survey of structures. *WIT Transactions on The Built Environment, 153*, 621–632.

Harp, D. R., Taha, M. R., & Ross, T. J. (2009). Genetic-fuzzy approach for modeling complex systems with an example application in masonry bond strength prediction. *Journal of Computing in Civil Engineering, 23*, 193–199.

Huang, J., Liu, W., & Sun, X. (2014). A pavement crack detection method combining 2D with 3D information based on Dempster-Shafer theory. *Computer-Aided Civil and Infrastructure Engineering, 29*, 299–313.

Ibarra-Castanedo, C., Sfarra, S., Klein, M., & Maldague, X. (2017). Solar loading thermography: Time-lapsed thermographic survey and advanced thermographic signal processing for the inspection of civil engineering and cultural heritage structures. *Infrared Physics & Technology, 82*, 56–74.

Koch, C., Georgieva, K., Kasireddy, V., Akinci, B., & Fieguth, P. (2015). A review on computer vision based defect detection and condition assessment of concrete and asphalt civil infrastructure. *Advanced Engineering Informatics, 29*, 196–210. https://doi.org/10.1016/j.aei.2015.01.008

Kumar, S. S., Wang, M., Abraham, D. M., Jahanshahi, M. R., Iseley, T., & Cheng, J. C. P. (2020). Deep learning-based automated detection of sewer defects in CCTV videos. *Journal of Computing in Civil Engineering, 34*, 4019047.

Lee, S. H., Han, J. H., Leem, Y. T., Yigitcanlar, T. (2008). Towards ubiquitous city: concept, planning, and experiences in the Republic of Korea. In *Knowledge-based urban development: Planning and applications in the information era.* (pp. 148–170). Igi Global.

Macher, H., Landes, T., Grussenmeyer, P., & Alby, E. (2014). Semi-automatic segmentation and modelling from point clouds towards historical building information modelling. In Ioannides M, Magnenat Thalmann N, Fink E, Zarnic R, Yen AY, & Quak E (Eds.) Digital heritage: Progress in cultural heritage: Documentation, preservation, and protection (pp. 111–120). Springer; Platz 3, D-14197 Berlin, Germany.

Mansuri, L. E., & Patel, D. A. (2020). Automated web-based condition survey system for heritage monuments using deep learning. In *The 5th International Conference on Sustainable Civil Engineering Structures and Construction Materials, Malaysia.*

Mansuri, L., Udeaja, C., Trillo, C., Kwasi, G., Patel, D., Jha, K., Makore, C.B., Gupta, S. (2019). Scientometric analysis and mapping of digital technologies used in cultural heritage field. In: *Association of Researchers in Construction Management, ARCOM 2019 - Proceedings of the 35th Annual Conference* (pp. 255–264).

Mansuri, L. E., & Patel, D. A. (2020a). Artificial intelligence-based automatic visual inspection system for built heritage. *Smart and Sustainable Built Environment in Press.* https://doi.org/10.1108/SASBE-09-2020-0139

Mohanty, S. P., Hughes, D. P., & Salathé, M. (2016). Using deep learning for image-based plant disease detection *7*, 1–10. https://doi.org/10.3389/fpls.2016.01419.

Monna, F., Magail, J., Rolland, T., Navarro, N., Wilczek, J., Gantulga, J.-O., Esin, Y., Granjon, L., Allard, A.-C., & Chateau-Smith, C. (2020). Machine learning for rapid mapping of archaeological structures made of dry stones–Example of burial monuments from the Khirgisuur culture, Mongolia. *Journal of Cultural Heritage, 43*, 118–128.

Nair, V., & Hinton, G. E. (2010). Rectified linear units improve restricted boltzmann machines. In *Proceedings of the 27th International Conference on Machine Learning (ICML-10)* (pp. 807–814).

Newman, C., Edwards, D., Martek, I., Lai, J., Thwala, W. D., Rillie, I. (2020). Industry 4.0 deployment in the construction industry: a bibliometric literature review and UK-based case study. *Smart and Sustainable Built Environment.*

O'Byrne, M., Schoefs, F., Ghosh, B., & Pakrashi, V. (2013). Texture analysis based damage detection of ageing infrastructural elements. *Computer-Aided Civil and Infrastructure Engineering, 28*, 162–177.

Ortega, J., Vasconcelos, G., Rodrigues, H., Correia, M., & Da Silva Miranda, T. F. (2019). Development of a numerical tool for the seismic vulnerability assessment of vernacular architecture. *Journal of Earthquake Engineering* 1–29.

Ouma, Y. O., & Hahn, M. (2017). Pothole detection on asphalt pavements from 2D-colour pothole images using fuzzy c-means clustering and morphological reconstruction. *Automation in Construction, 83*, 196–211. https://doi.org/10.1016/j.autcon.2017.08.017

Parveen, S. (2014). Surat: As a major port town of Gujarat and its trade history. *IOSR Journal of Humanities and Social Science, 19*, 69–73.

Pérez-Gracia, V., Caselles, J. O., Clapés, J., Martinez, G., & Osorio, R. (2013). Non-destructive analysis in cultural heritage buildings: Evaluating the Mallorca cathedral supporting structures. *NDT E International, 59*, 40–47.

Petti, L., Trillo, C., & Makore, B. N. (2020). Cultural heritage and sustainable development targets: A possible harmonisation? *Insights from the European Perspective Sustainability, 12*, 926.

Pieraccini, M., Fratini, M., Parrini, F., Pinelli, G., & Atzeni, C. (2005). Dynamic survey of architectural heritage by high-speed microwave interferometry. *IEEE Geoscience and Remote Sensing Letters, 2*, 28–30.

Ramcharan, A., Baranowski, K., Mccloskey, P., Ahmed, B., Legg, J., & Hughes, D. P. (2017). Deep Learning for Image-Based Cassava Disease Detection. 8, 1–7. https://doi.org/10.3389/fpls.2017.01852.

Ramos, L. F., Miranda, T., Mishra, M., Fernandes, F. M., & Manning, E. (2015). A Bayesian approach for NDT data fusion: The Saint Torcato church case study. *Engineering Structures, 84*, 120–129.

Ren, S., He, K., Girshick, R., & Sun, J. (2015). Faster r-cnn: Towards real-time object detection with region proposal networks. In *Advances in neural information processing systems* (pp. 91–99).

Reyno, T., Marsden, C., & Wowk, D. (2018). Surface damage evaluation of honeycomb sandwich aircraft panels using 3D scanning technology. *NDT E International, 97*, 11–19.

Riggio, M., D'Ayala, D., Parisi, M. A., & Tardini, C. (2018). Assessment of heritage timber structures: Review of standards, guidelines and procedures. *Journal of Cultural Heritage, 31*, 220–235.

Saadi, A., & Belhadef, H. (2020). Deep neural networks for Arabic information extraction. *Smart and Sustainable Built Environment.*

Scarre, C., & Roberts, J. (2005). The English cemetery at Surat: Pre-colonial cultural encounters in western India. *The Antiquaries Journal, 85*, 251–291.

Taboroff, J. (2000). Cultural heritage and natural disasters: Incentives for risk management and mitigation. *Managing Disaster Risk in Emerging Economies, 2*, 71–79.

Tzutalin: LabelImg (2018). Retrieved December 3, 2018, from https://github.com/tzutalin/labelImg.

Udeaja, C., Trillo, C., Awuah, K. G. B., Makore, B. C. N., Patel, D. A., Mansuri, L. E., & Jha, K. N. (2020). Urban heritage conservation and Rapid urbanization: Insights from Surat India. *Sustainability, 12*, 2172. https://doi.org/10.3390/su12062172

Valero, E., Forster, A., Bosché, F., Hyslop, E., Wilson, L., & Turmel, A. (2019). Automated defect detection and classification in ashlar masonry walls using machine learning. *Automation in Construction, 106*, 102846.

Wang, N., Zhao, X., Zhao, P., Zhang, Y., Zou, Z., & Ou, J. (2019). Automatic damage detection of historic masonry buildings based on mobile deep learning. *Automation in Construction, 103*, 53–66. https://doi.org/10.1016/j.autcon.2019.03.003

Yeum, C. M., & Dyke, S. J. (2015). Vision-based automated crack detection for bridge inspection. *Computer-Aided Civil and Infrastructure Engineering, 30*, 759–770.

Zou, Z., Zhao, X., Zhao, P., Qi, F., & Wang, N. (2019). CNN-based statistics and location estimation of missing components in routine inspection of historic buildings. *Journal of Cultural Heritage.* https://doi.org/10.1016/j.culher.2019.02.002

Chapter 2
Smart Cities, Grids, Homes and the Workforce: Challenges and Prospects

Abouzar Estebsari and Edmundo Werna

Abstract The chapter explores specific developments and opportunities in smart technology in cities as well as the interface with the workforce. The first part of the chapter looks at technological development in smart grids and smart homes: integration of new Information and Communication Technologies (ICT) such as Internet of Things and advanced algorithms based on Artificial Intelligence (AI) to manage energy in urban areas and support user awareness or operator observation. The "smartness" of smart cities is defined. The enablers including AI and ICT, and their applications and benefits for users are discussed. The interoperability layers of smart grids architecture model are presented, and the importance of multidisciplinary ex-ante analysis of newly developed technologies and distributed intelligence is highlighted. The second part relates to the urban workforce. This is distinctive as most literature on smart cities looks at the impact on people as residents/users—not as workers. This part starts by discussing the overall impact of the development of smart grids and homes in the workforce across the sectors of the urban economy. Following, it concentrates on workers in trades related to the built environment. It argues that technological developments need a restructuring of vocational training, which may lead to a more skilled workforce with better working conditions. At the same time, it may reduce the quantity of employment, therefore part of the wealth generated by economic growth derived from technological advancements should be used to strengthen social protection. The chapter makes proposals to address current challenges, reaching a win–win situation.

1 Introduction

There are several projects across the world to develop smart cities and systematically shift them from digitalisation toward Ubiquitous Cities. London Smart Vision (2013), Amsterdam Smart City (2019), Singapore Smart Nation (https://www.smartnation.gov.sg) or San Francisco Environment Project (2011) are

A. Estebsari (✉) · E. Werna
London South Bank University, London, UK
e-mail: estebsaa@lsbu.ac.uk

R. T. M. Li et al. (eds.), *Current State of Art in Artificial Intelligence and Ubiquitous Cities*, https://doi.org/10.1007/978-981-19-0737-1_2

some examples of such initiatives. Ubiquitous Cities integrate novel technologies and advanced methodologies to distribute intelligence across the society. The key applications include smart transportation, smart public health, smart environment, smart grids and smart buildings. Digitalization is the first stage of creating ubiquitous cities. Anthopoulos and Fitsilis in (2010) review several digitalised cities in their early stages. Kyoto and Amsterdam started using web and Virtual Reality for managing streets, malls, etc. (Ishida, 2002; Van Lieshout, 2001). Copenhagen in Denmark used public database to cover local needs (Bastelaer, 1998). Smart city of Brisbane in Australia added AI to the decision-making services. And in Digital City of Craigmillar in Scotland, groups of citizens shared knowledge and social services (Bastelaer, 1998).

As one of the main objectives of smartening the cities is sustainability improvement and green modernisation, the focus of this chapter is on the energy sector including distribution and consumption premises. The two application domains, smart grids and smart homes, are discussed from two aspects: technological aspect with a review of enablers and their applications, and workforce aspect including the workers' benefits and challenges.

Smart grids and homes integrate novel Information and Communication Technologies (ICT) with bidirectional data exchange capability and embedded intelligent algorithms. To develop ideas and improve the smartness of these systems, standards are essentials. Standards, in general, improve the quality of life, encourage innovation and competition, and promote international trades. There is a high demand to develop frameworks for the standards. In smart grids, there have been several developed frameworks to define standards, including NIST (National Institute of Standards and Technology) (NIST, 2016), Smart Grid Interoperability Panel (SGIP) (NIST, 2012), or IntelliGrid Architecture sponsored by the Electricity Innovation Institute (E2I) as an ICT Program (Hughes, 2005), in the USA, Smart Grid Architecture Model (SGAM) in the EU (Estebsari et al., 2019), the State Grid Corporation of China (SGCC) Smart Gid Framework in China (n.d.), and Japanese Industrial Standards Committee (JISC) in Japan (www.jisc.go.jp, n.d.). In these frameworks, a set of descriptions are provided to make a basis for analysing the assets, use cases, interfaces, and standards of smart grids. Such reference models allow for shifting existing paradigms to "smart" systems systematically through maximising reuse of the in-place assets and off-the-shelf technologies.

Understanding the enablers of smart grids and homes allows for developing new use cases. And mapping the newly developed use cases to reference models help identify the gaps and analyse the interoperability of different technologies to deploy the new devices or schemes more efficiently. This would support for reviewing the benefits and challenges of new deployments for the users, especially the workers in the smart society. Therefore, in this chapter, the first section introduces the evolution enablers of smart grids and smart homes with some applications. Then, the impacts on the workforce in urban societies are discussed in the second section.

In the first section, two common enablers including Artificial Intelligence (AI) as methodological enabler and ICT as technological enabler are reviewed. The key applications of these enablers to smart grids and homes are discussed with some examples. Internet of Things (IoT) solution is introduced as an ICT-based approach to facilitate creation of ubiquitous cities with a focus on smart grids and homes. The direct benefits of emerging technologies in smart grids and homes on human, especially the operators and workers, are presented with some illustrative examples. These benefits are discussed in twofold: advantages of new ICT-enabled approaches to support ex-ante safe test and verification of new use cases, and benefits of ICT and AI in improving operators' efficiency and satisfaction.

Section 2 is based on key research by academics and institutions, notably the comprehensive studies of WBCSD (2020) and Goger et al. (2019), the presentation of Yuson (a key leader of the construction workers) (Yuson, 2020) and related references. The section is also informed by the experience of one of the authors (Werna) during 16 years at the International Labour Office including the opportunities to discuss the future of work with workers, employers and governments, to prepare and take part on meetings and to implement projects on the ground, also informed by the literature quoted herewith.

2 Smartness in Ubiquitous Cities

Prior to any discussions on the so-called smart society, smart grids, smart home, etc. the definition of "smartness" and the reason such environments are called "smart" should be noted. According to the Cambridge Dictionary, "smart" means intelligent or able to think quickly or intelligently in difficult situations (dictionary.cambridge.org., n.d.). Any device or system which is capable of independent and seemingly intelligent actions is called "smart" (Cambridge.org., 2019). The question is how a society or system becomes smart. There are two means to make them smart: (a) enabling them to communicate bidirectionally using sensors and actuators, and (b) through embedding elaboration systems to run mathematical computations and logical operations to make decisions. For example, in smart grids, the digital technology allows for two-way communication between the utility and its customers, sensing along the transmission lines, and actuating physical elements (Ghatikar, 2016; Siano, 2014).

In this section, the two main technical drivers of smart cities including technological and methodological enablers are firstly introduced as "evolution enablers". The focus of this chapter is on smart grids and smart homes. The services and applications are elaborated with some examples for novel ICT schemes as technological enablers and AI as methodological enablers. Regarding the ICT-based schemes, Internet of Things applications in smart grids and homes are briefly addressed as well. This part is entitled "application of enablers". The section will conclude with reviewing some opportunities that these technologies and methodologies would bring to the system operators and users. This part is presented as "direct benefits for human". The impacts on the urban workforce will be then detailed with challenges and prospects in the second section of the chapter.

2.1 Smart Grids and Smart Homes: Evolution Enablers

Smart grid is defined as electricity networks that can efficiently integrate the behaviour and actions of all users connected to it to ensure an economically efficient, sustainable power system with low losses and high quality and security of supply (Force & SFTF, 2009). The drivers toward smart grids, however, are not limited to technological drivers. There are public policy drivers including restructuring, deregulations and national security, as well as market drivers including consumer demands, privatisation and competition (Borlase, 2017).

The smart revolution to form intelligent cities requires two important enablers: technologies and methodologies. The enabling technologies cover a wide range of devices, software, systems, function and standards to digitalize the cities and reshape control and management of societies. The technologies include physical devices like sensors and actuators which are widely distributed, exchanged information management through signal processing, efficient and reliable communication networks, and data analytics and distribution. The methodologies add smart functions to better manage the digitalized systems. They include programming and simulation techniques and tools, optimization methods, and system modelling (Fig. 1). In this chapter, Information and Communication Technology (ICT) from the technologies, and Artificial Intelligence (AI) from the methodologies are elaborated with application focus in smart grids and homes (Kumar et al., 2020).

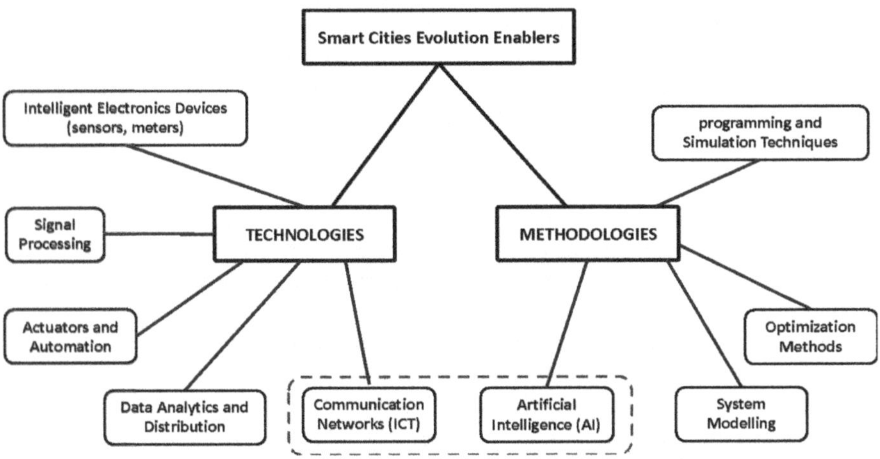

Fig. 1 Smart Cities Evolution: technological and methodological enablers

2.2 Smart Grids and Smart Homes: Applications of Enablers

There is a wide range of technologies and methodologies which enable evolution of smart grids and smart homes, but the two enablers are more common and vital nowadays: Artificial Intelligence (AI) and Information and Communication Technologies (ICT). The former is more on *methodologies*, while the latter brings novel *technologies* to establish ubiquitous cities.

AI Applications: Growing capability of computation machines allows for practicing and improving some old data and operational methods to develop more efficient tools. Machine learning (ML) algorithms are among those methods whose basis could be found in older scientific and engineering literature, however the implementation of most of those methods required high performance computation (HPC). Today's opportunity of advanced HPC brings them to life and support developing more advanced techniques. For example, most of Deep Learning algorithms are based on multi-layered Neural Networks with more nodes and layers which increase the accuracy of the outcome. For example, (Hajiabadi et al., 2020) uses deep learning with loss ensembles for solar power prediction in smart cities.

Load forecasting has improved a lot in the recent decades for the accuracy and performance of the techniques, but new smart city applications demand for more advanced methods. For instance, emerging smart homes require smart energy management, which implies embedding technologies and algorithms exchanging data through home communication network (HAN). Applying the same forecast tools used in power grids would not capture the fast dynamics of volatile loads of a single residential household, hence new methods need to be developed. (Estebsari & Rajabi, 2020) presents a novel model based on deep learning and image encoding techniques to forecast loads in single residential homes.

Bringing smartness to individual assets of the society such as energy components, homes, vehicles, and people accessories mean coordination and management of a large amount of data. This is challenging even if the communication networks grow, as processing and analysing such big data would demand very high-performance computation with advanced AI. Therefore, in the *transition toward ubiquitous cities*, digitalization is not sufficient, and the intelligence should be distributed (Pau et al., 2018).

There are many local or small-scale AI-based controllers and agents which are designed and deployed for specific applications with limited amount of data. User awareness of energy consumption and generation at home is an example of applying AI in smart homes. New algorithms can be embedded in modern smart meters to support home energy management, reduce costs and emission, and improve user comfort. Demand Side Response (DSR) algorithms (Barbierato et al., 2019; Conejo et al., 2010), Non-intrusive appliance load monitoring (NIALM) (Asres et al., 2021), battery management systems (Bompard, et al., 2016), lighting and thermal control (Oh et al., 2020) are some examples of enablers to which AI would add values.

ICT Enablers: ICT related systems are technological enablers to deploy *ubiquitous* city projects. In this subsection, the role of ICT in smart grids with respect to the traditional power systems is reviewed as an example of technology applications to smart cities.

The common terms in most of smart grid definitions are efficiency integration of actions, incorporation of digital technologies, information exchange and two-way communication technologies. Therefore, the communication is a key element in smart grids. This is not only the glue to interconnect physical devices, but also market actors, operation centres, enterprise entities, energy producers and users, and other business premises. The ICT layer covers *(i) information to be exchanged, (ii) communication protocols and standards, (iii) media and (iv) control or supervisory systems.* In electricity utilisation premises the protocols like IEEE 802.11, IPv4, IPv6, IEC 61,850, the media like ZigBee, HomePlug, Z-Wave (HAN), and the control or supervisory systems such as Building Area Network (BAN), OpenADR (price responsive) and energy management system are used.

In distribution domain, the ICT allows for introducing many new services that enhance energy network operation and efficiency. It supports distribution management systems in 3 parts: data acquisition, system modelling, and system operation.

Data acquisition: ICT interconnects many Remote Terminal Units (RTUs) distributed across the network and buildings with centralised or decentralised data gateways such as Supervisory Control And Data Acquisition (SCADA) system (Upadhyay & Sampalli, 2020). Device status, energy data and user information are periodically exchanged thanks to well-established ICT network. Any data storage, event log and reporting are performed in this part. The ICT supports smart payment of bills, customer complaint handling and outage affected customer identification (Chiemezie et al., 2013; Järventausta et al., 2008).

Modelling: ICT enables modelling applications such as state estimation (Pau et al., 2019), fault analysis (Estebsari, 2016), route optimization for field crew and power flow calculations. This is through providing prompt information and distributing the intelligence to enhance computation power and modelling performances.

System operation: the existing services would be automatized and enhanced when ICT is integrated. Network real-time monitoring of distributed renewable generation, user energy consumption and topology reconfiguration are facilitated by appropriate deployment of an ICT-enabled system (Kou et al., 2020; Mohseni et al., 2021). In this application, smart metering infrastructure like Advanced Metering Infrastructure (AMI) utilizes ICT and widely distributed smart meters to improve monitoring. Protection coordination, network reconfiguration, power loss minimisation, automated mapping, fault and outage management (Bahmanyar et al., 2017), and operator training are some of other examples of the ICT-enabled system operation (Prostejovsky et al., 2019).

Internet of Things: IoT-based approaches are among popular and common novel ICT-based enablers. IoT aims to interconnect numerous heterogeneous devices through the internet. This is performed through a layered architecture in which the devices, the people and cloud-based services are integrated to support an application (Khajenasiri et al., 2017). IoT outperforms other ICT based approaches especially in use cases where large number of distributed devices should communicate bidirectionally. Smart grids require a lot of services that involve numerous widely distributed devices, such as smart meters. In an urban area, there may be over millions smart meters both in customer premises and field substations. They should periodically send signals to report alarms, status or measurements, and receive signals of actuation commands. The signals may be used by several services, for instance fault location (Estebsari, 2016), thief detection, energy balance, power quality monitoring, demand side response (Dyson et al., 2014) and state estimation (Pau et al., 2019). IoT enables interactions between the same devices with many services.

The performance of IoT-enabled schemes for smart grids depends on several parameters including the number of deployed intelligent electronic devices (IED) in the network, the cloud protocols (e.g. MQTT, REST, etc.), interoperability, and fast bidirectional communication (Pau et al., 2018). Smart meters are low-cost multi-purpose IEDs that can be widely utilized by an IoT-based architecture. The architecture collects real-time data from large number of different meters and sensors to provide actual information for the services. The services should have appropriate Application Programming Interface (API) to access data. This allows for integrating novel services and distributed applications with minimum effort and a plug-and-play fashion. It is important to ensure interoperability among different existing communication protocols. This requires some middleware to abstract hardware functionalities from different low-level technologies through uniform interfaces. For some services, near real-time data exchange is needed that implies an asynchronous communication. This is often implemented by using the publish/subscribe approach (Eugster et al., 2003).

2.3 *Smart Grids and Smart Homes: Direct Benefits for Users*

The benefits and opportunities provided by smart grids and smart homes can be reviewed from different aspects. In this section, the direct benefits of integrating new technologies and methodologies for the system operators and users in smart grids and homes are discussed. In the next section, the benefits and challenges of this evolution for the urban workforce and built environment workers are elaborated in detail.

Improving ex-ante interoperability analysis: software and hardware in-the-loop

Ex-ante analysis of newly developed technologies and methodologies is crucial to verify the performance and efficiency of them (Bottaccioli et al., 2017). This is normally done by modelling and simulation; however, the multidisciplinary nature of modern systems requires new sophisticated tools and methods for reproducing the behaviour of the real system in the lab (Stevic et al., 2017).

In each application domain of ubiquitous cities including smart grids and homes, the actors and actions which form use cases may belong to different disciplines. For example, in smart home applications, users and devices are some of the actors and home energy management is an action; users' behaviour is a social output which is an input to define the function of appliances and their status, the appliances are electrical or thermal physical devices, and the home energy management is a smart function integrating ICT and AI. Multidisciplinary systems like this example, multi-vector energy systems like interconnected gas and electricity networks (Mazza et al., 2019), and multi-level hierarchical systems like utilisation-distribution-transmission networks (Stevic et al., 2016), imply the fact that multidisciplinary modelling and analysis is needed. The new modelling approaches should be able to study the interoperability layers—including physical, information, communication, function and business layers (Fig. 2 (CEN-CENELEC-ETSI, 2012)) of smart grids and homes. Reviewing literature and project reports shows a growing interest in developing new schemes which are ubiquitous, general purpose, scalable and flexible. The new management or operation schemes integrate different modules from different disciplines. Most of recent investigations and implementations proved Internet of Things (IoT) to be an efficient and low-cost approach to improve the ex-ante modelling and analysis of smart grids.

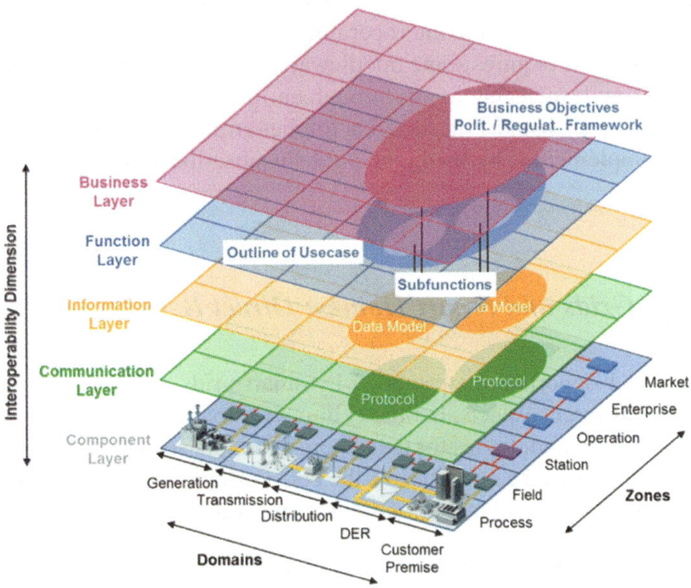

Fig. 2 Smart grid architecture model (SGAM) with interoperability layers (CEN-CENELEC-ETSI, 2012)

Improving real-time efficiency of system operators: human in-the-loop

All technological and methodological enablers aim at evolving smart grids and smart homes through digitalization and automatization. Nevertheless, human would remain in the loop. Not only in planning phase, but also in operation the technicians, engineers, regulators, policy and decision makers are closely involved in running the systems. In the following, some examples of the role of ICT in increasing the efficiency of human operation in smart grids are introduced.

SCADA in electricity distribution networks provide appropriate data from preprocessed analysis and classification of events to the operators via Human–Machine Interface (HMI). Measuring consumption or generation is performed remotely and the bills are delivered via various methods automatically. New systems provide facilities for the customer service operators to log the complaints and map the customers. In case of scheduled maintenance, the customers would be notified in proper time.

Route optimization for technical crew who are dispatched to field is another example of ICT and AI application for human satisfaction. The terrain conditions and information of the streets reaching the filed are stored in database to be used in transportation models aiming at optimizing crew travel. In some restoration or repair missions, the time is crucial as it impacts on the reliability of the distribution system (Bahmanyar et al., 2016; Estebsari et al., 2018).

Most of electricity outages are due to distribution network fault. When a fault in network disconnects the circuit or triggers the protection relays to interrupt the electricity, the faulted lines are isolated to restore the rest of system. This process traditionally should take around an hour (40–80 min) in an urban area. Figure 3a summarizes the whole restoration process in a conventional outage management system. In this process, the human (the electrical engineer or technician) should firstly localize the outage based on trouble calls or breaker alarms. The repair crew is dispatched to travel to the field. They need patrolling and investigating the problem, and finally isolate the fault. In practice, this is a challenging and time-consuming process. There may false or fake reports, patrolling is demanding specially during

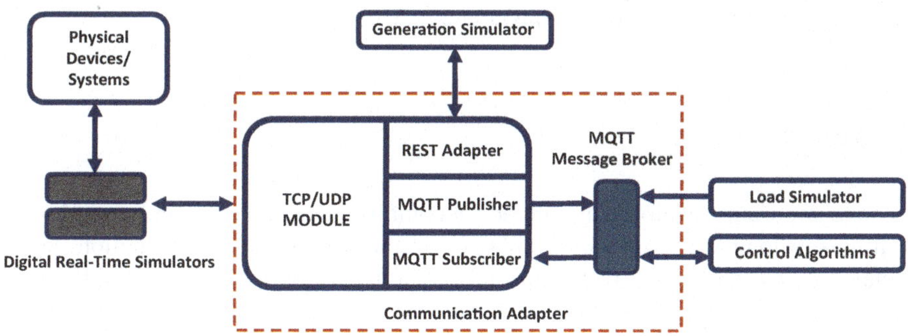

Fig. 3 An IoT-based proposed scheme for real-time co-simulation in smart grids (Bottaccioli et al., 2017)

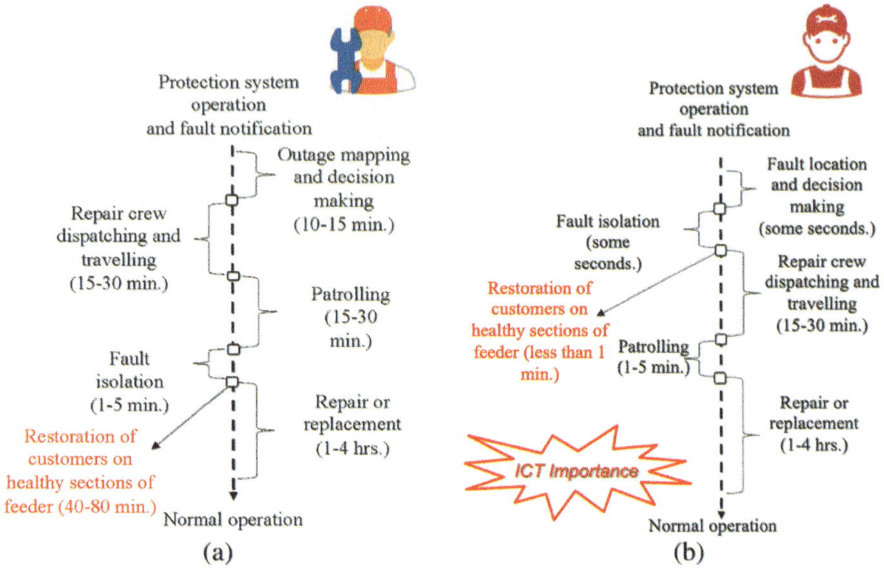

Fig. 4 Example of technician efficiency improvement using AI and ICT in smart grids (Bahmanyar et al., 2016)

adverse weather conditions, and the night darkness or fewer reports make the outage identification longer.

AI can support localizing faults and outage faster (Bahmanyar et al., 2017), Intelligent Electronic Devices (IED) can provide data over communication system, and automatized actuators can receive commands to isolate the fault before human arrives. Figure 3b shows the importance of ICT in reducing the restoration time to less than a minute. This would significantly improve system reliability (Falahati et al., 2012; Ren et al., 2019) and human satisfaction at both operator and customer sides (Bahmanyar et al., 2016). Figure 4 shows an example of technician efficiency improvement using AI and ICT in smart grids (Bahmanyar et al., 2016).

3 Smart Labour

Despite important developments already accomplished, smart grids and smart homes still have a long way to go. As explained in the abstract and introduction, urban technology has a meaning if it contributes to human development—as residents but at the same time as workers. One needs to have a decent income to live decently in a city or town. The actual impact on the workforce is still a subject of inquiry, as the technology is still unfolding. The title of the section is an allusion to the overall use of the word 'smart' (as applied in cities, homes and grids).

This section will discuss trends, potential consequences and provide insights into the future.

First, the session discusses the overall impact of the development of smart grids and homes in the urban workforce in general (i.e. across the sectors of the urban economy). Following, it concentrates on workers in trades related to the built environment—i.e. the sector of the economy which encompasses the grids and homes. In each case, the section will discuss existing and potential opportunities and challenges, paving the ground for conclusions and recommendations.

Urban workforce

Opportunities

Starting with the broad picture. Smart grids accelerate communication and the transmission of energy, which, in turn, pave the ground for a more efficient economy in the localities where they are placed. When the economy is more efficient and grows, it has more potential to generate new employment and to improve job quality. From this perspective, smart grids are potentially good news for the workers in such areas. Naturally one cannot guarantee that the economy will indeed benefit workers, as there are cases in which surplus is unevenly distributed, sometimes to the extreme. But at least a growing economy has more prospects than a stagnated or shrinking one.

Faster and better communication has a great impact on online working. It opens many possibilities. Those whose work can be done online are able to offer their services to clients literally all over the world. Those who have limited mobility are particularly benefited—e.g. workers with disability, those who live in neighbourhoods badly served by public transport, those who live in far away suburbs.

WBCSD (2020), in its research about future trends in work, noted that where once physical proximity was required for people to get work done, the advancements of digital communication technologies and collaboration platforms have allowed for an accelerating trend of remote working—a phenomenon known as telecommuting—and created the opportunity for more distributed teams.

According to a survey conducted by IWG (2019), over half of professionals worldwide already work remotely at least 2.5 days per week. Moreover, in the past two years alone, there has been a 78% increase in LinkedIn job posts advertising flexible work arrangement. These statistics indicate that remote work is here to stay and is not only a temporary trend in the labour market (Felstead & Henseke, 2017; Sako, 2021; Remote Year, 2019).

Faster electricity contributes to the development of the gig economy, immersive tech, IoT. IoT is also instrumental for the grids themselves. They will be analysed in turns. The gig economy is defined as (WBCSD, 2020):

… "a labor market that brings together supply and demand for a set tasks or projects… Distinct from times when workers and employers had to find each other through job adverts and recruitment agents, so-called online gig workers are onboarded by independent platform businesses and then matched to work opportunities by the platform, who typically also handles the rate, service delivery standards, cash collection and payment" ...

Many gig workers want to work in this way. Self-perceived benefits include the flexibility, lack of monopolization of time, the earnings potential, and the lack of management overhead. For many, a gig is not a career, but a 'side hustle', an opportunity to generate revenue (Dolber et al., 2021).

Virtual reality (VR), augmented reality (AR) and mixed reality (MR)—three immersive technologies that fall under the category of extended reality (XR)—have been expanding throughout the past decade. In recent years, their presence began to be increasingly felt beyond the fields of gaming and entertainment, where they initially gained popularity. According to a global survey by WEF (Sako, 2021; World Economic Forum, 2018), more than half (58%) of companies are likely to adopt VR and AR by 2022.

There are already some interesting case studies in the use of immersive technology in the world of work. In job advertising (Jaguar Land Rover/Gorrilaz), in job assessment (Lloyds Banking Group) in job training (Walmart Black Friday simulator), in employer branding (WeWorkAR video) and in even in the creation of an entirely virtual organization (eXpRealty virtual office island).

"The Internet of things (IoT) is a network of physical objects—devices, vehicles, appliances—embedded with sensors, software, network connectivity and computing capability that enable them to collect, exchange and act on data, usually without human intervention" (WBCSD, 2020). It accelerates the economy. It will be beneficial for those who have access to it, as it increases efficiency. It is not clear if it will increase or decrease employment. For sure, it will need re-skilling (such as the ones above).

Homes

Moving from smart grids to smart homes, the latter also offer a number of existing or potential opportunities to workers. Overall, it saves time for domestic chores. Such time can be used for gainful work, or to rest.

There is ample evidence that home-based enterprises have provided livelihoods for large numbers of workers, both in the Global South and Global North. A large number of people are not able to find employment and have to resort to self-employment. Out of these, several work from home for different reasons. It is not only the sophisticated online workers, able to tele-work. It is also a large number of self-employed blue-collar workers or entrepreneurs who cannot afford to open an office elsewhere, and/or—especially in the case of women—have to combine work with home/children care. Many also want to spend more time with family. For those who are or plan to work at home, smart homes provide an added value as some of the appliances may be used for work as well. Or, as said before, save time. This is in normal conditions. When crises strike, such as COVID19 has shown the world, work from home for many people turn out to be the only option—and therefore smart homes become even more important.

3.1 Challenges

To recap, smart electricity contributes to the adoption and/or development of the gig economy, immersive tech, IoT. It also contributes to 3/4 D printing and blockchain (the latter is a key connector across a massive flow of data list of evolving records which use sophisticated cryptography to link back to each other and contain transactional data allowing market participants to execute transactions without the need for a centralized third party—i.e. a bank (WBCSD, 2020). These developments, together with smart homes, bring challenges to workers. One cannot lay the blame on smart grids and/or smart homes for all the challenges noted below. The point is that such challenges are connected to technological developments which are facilitated by smart grids/homes, therefore they are mentioned here.

ILO (2019) and WBCSD (2020) also noted that the unprecedented pace of development and adoption of such new technologies are raising concerns of mass job displacement and job losses while rapidly changing tasks and work patterns and challenging how companies and commerce function.

While such technologies may be beneficial for the workers who use them (Sect. 2.1.), there are overall challenges for the urban workers, including increase in inequality, increase in casualization of work, lack of social dialogue, job losses and lack of socialization. They will be analized in turns.

The first one is the potential increase in inequality between those who have access to such technologies and those who do not. By and large, industrialized countries are well ahead of developing ones in technological innovation. Considering the level of poverty specially (but not only) in developing countries, but the race for the new technologies is also likely to increase the inequality gap, unless policies to help those who do not have access are put in place.

Second, the types of professional work aligned with the aforementioned technological advancements tend to be temporary or project-based. Workers are hired to complete a particular task or for a certain period of time. This means that workers have to be constantly on the look-out for the assignments, and there is no job security. While this was already been escalating since the 1970s, the power of the new technological developments can take it to a much greater level (WBCSD, 2020).

Third, social dialogue with employers and governments has traditionally been a powerful means for workers to collectively bargain for better wages and working conditions (Costa, 2021; Van Empel & Werna, 2008). However, in such type of work arrangements explained above, social dialogue is very difficult or almost impossible. Contrary to traditional industrial factories, where there was a clear employer, in the new arrangements it is difficult or impossible to determine who and where is the employer. This makes it practically impossible to establish social dialogue to improve labour conditions. As an anecdote: the title of an internal ILO study on the mining sector in Turkey was "*Who is the Employer*"? The title says it all. The structure of the sector was so intricate with layers of sub-contracting and several modalities of contract, that in the end it was not clear who would be responsible if an accident happened.

The fact that workers under such arrangements are far from each other also conspire—at least in a first instance—against social dialogue. In other words, when workers are physically in the same worksite and know each other, it is much easier for them to discuss collectively and organize themselves to dialogue at least with the government—and with the employer if there is one that can be identified and located. There is less synergy among workers who work online, especially those who do carry out collective assignments. When a given group of workers is spread among different countries, it is difficult to find a common ground to dialogue with the government.

Having said the above, it is also important to note that workers may benefit from ICT to speed up exchange of information, online dialogue among them and decision-making. It may be more realistic to explore potential benefits of technological development rather than to fight them.

The fourth point is that such technologies can lead to job losses for being (highly) capital-intensive. This links to the traditional debate regarding labour- versus capital-intensive techniques (Salazar-Xirinachs et al., 2014). Labour-intensive techniques may create more jobs during a certain period as opposed to capital-intensive ones, but it is regarded as slowing down economic growth. If capital-intensive techniques have a sound effect on economic growth, in the end it will create more jobs in absolute terms. For example, while a house built with 3D printing will cut down jobs needed to build it in a traditional way, the massive manufacture of 3D printing and the consequent massive production of housing in such a way may in the end generate more jobs. It is difficult to make a conclusive prediction. But one way or another, given the impact of industrial growth in the planet, it is important to bear in mind that economic growth is finite, and one cannot rely on it forever to generate jobs via ever-expanding capital-intensive techniques. This is an important debate, that must continue to bring new findings.

During a few decades, industrialized countries have set up production unit in developing countries. Despite of many debatable practices, this has created many jobs in the latter set of countries. By and large, industralized countries are well ahead of developing ones in skills training for technological innovation. The race to digitalization may lead to a counter shift. I.e. jobs migrating back to the North, a fact which may exacerbate unemployment in the South (Yuson, 2020).

If or where job losses take place, unemployment benefits are needed. But unemployment also have other consequences, even if the unemployed has financial compensation. Having a professional activity is an important element in one's insertion in society. There is evidence of mental health problems (such depression) derived from unemployment (Batic-Mujanovic et al., 2017).

The fifth challenge noted here concerns both the unemployed and those who work online. Socialization is one of the important elements of collective work. To work in isolation (or not to work) has a toll on one's health (Werna et al., forthcoming). Also, according to WBCSD (2020), disconnected workplaces have been linked to nearly 40% higher rates of absenteeism, 50% higher rates of bodily accidents and 60% higher rates of avoidable errors. Further studies have shown that a connected and engaged workplace fosters higher rates of self-reported job satisfaction, improved job performance and increase rates of retention.

As noted by Werna and Klink (Mella & Savage, 2017), a dystopian scenario would entail masses of unemployed workers. Unless the wealth generated by those who work would be evenly distributed, there will be alarming problems. The ethical and social toll would entail an increase in poverty, riots and crime. In terms of the economy, one can ask: who will buy the products made by machines, if nobody has cash?

Above are the challenges to urban workers brought up by technological developments, which are facilitated by smart grids and/or brought up by smart homes. One issue that cuts across most of the challenges is disaggregation: workers in a same project but in different locations; the end of the physical workplace as we know it; production, exchange and purchase of goods without the need for human contact. This changes the perspective not only of urban work specifically, but also the concept of the 'urban' itself. A basic distinctiveness of cities and towns is gathering, and economies of agglomeration is one of its crucial features. Therefore, if economic activity does not need agglomeration any more, this may have a significant impact in urban morphology.

Those who are able to work from online (especially in smart homes, or even elsewhere), may opt to live in the countryside or small villages, which is already happening to some extent. This is still incipient if one takes into account the number of settlements and people in the planet, but is a trend or a desire at least for part of the urban workers. COVID19 came to exacerbate the dispersion trend.

There are still many people who like to live in urban areas because of the social and cultural life, which is a counter-trend. We do not know yet which trend will prevail. So far there has been a steady and constant growth towards urbanization. It is important to monitor this, think about the possible consequences for each one of the alternatives and how to best design the human settlements of the future.

3.2 *Built Environment Workers*

The built environment encompasses the very trades which produce grids and homes. Therefore, out of all urban workers, those in trades related to the built environment are particularly impacted by the technological innovations noted in Sect. 1 of this chapter.

<u>Opportunities</u>

BWI (Building and Wood Workers International) is the one and only international confederation of trade unions in construction-related (and forestry-related) trades. Mr. Ambet Yuson, BWI's Secretary General, is optimistic in relation to advancements in artificial intelligence, also informed by a study commissioned by BWI (Goger et al., 2019). In a recorded interview (Yuson, 2020), the Secretary-General notes that such technological developments are welcome. They are likely to be safer, provide new training and new working opportunities specially for women and youth, pay better, need less working hours (meaning more time for the family) and have more sustainable production methods.

The points made by Yuson (2020) and Goger et al. (2019) are valid at least as premises and relate to (i.e. address) key current challenges of construction workers. They will be analysed in turns.

Construction is one of the most dangerous occupations. Approximately 17 per cent of fatal accidents reported globally take place in the construction sector, amounting to 60,000 fatal accidents per year. Worldwide, around 22 construction workers die annually for every 100,000 construction workers. The construction industry accident fatality rate stands at more than double that of the 'all-sector' average. It is estimated that fatality rates in the least developed countries may be more than double the rates in developed economies (International Labour Office, 2015; Lawrence & Werna, 2009; Mella & Savage, 2017).

A substantial part of the safety and health deficits is due to the hazardous nature of the construction work. This is likely to improve with digitalization in general, including smart grids and smart homes. The nature of the (so-called) 'smart' work is less hazardous.

As smart grids and smart homes entail technological innovations in many places around the world, they will bring new training and job opportunities for workers. The specific groups signaled by the Secretary General of the Building and Wook Workers International (the trade union confederation for construction) (Yuson, 2020) (youth and women) are indeed priority beneficiaries. First, there is ample evidence about the difficulties for a large number of young people to enter the labour market, with grave forecasts for the increase in poverty, alienation and insecurity (Werna, 2008). Smart grids and homes provide an opportunity for the youth. Senior workers do not have an advantage because, as the technologies are novel, they do not have the professional knowledge already. In addition, young people usually deal better with digital innovations than older generations. On the other hand, this may also create challenges for senior workers, which will be discussed later.

In regard to women, according to the latest publication of the ILO on labour in construction (International Labour Office, 2015), despite some increases, the participation of women in such sector remains low, and lower than in many sectors. Equal opportunities have not been achieved. Women are sometimes employed as part of a family work unit, often without receiving direct payment. The percentage of all women paid workers in construction—which denotes the share of paid women working in the construction sector, out of female workers in all sectors—is between 5.5 and 14.6 percent according to the geographical region. The share of women workers out of all workers in the construction industry is between 0.5 and 7.5 percent according to the region.

At least a significant part of the exclusion of women is due to the traditional view of construction as a sector which requires heavy physical work, regarded as task for men. This has created an entrenched culture. As the type of work that smart grids and homes require is more sophisticated than traditional "heavy load carrying" construction, it helps to counteract this routed view.

The next point is better salaries. Construction is a pyramidal sector, with a large number of low-paid workers at the bottom. Moreover, although payments below what the law stipulates and/or longer working hours without proper compensation are illegal, such practices do happen given the high numbers of informal or casual workers, and/or lack of control. Delays in payment are also an extended practice (Buckley, 2018; Wells & Prado, 2019). Given the innovativeness and sophistication of smart grids and smart homes, their implementation will need highly skilled workers, who are likely to be better paid. The impact on the unskilled and semi-skilled workers remain uncertain, and will depend on a number of variables (such as the capacity of the workers to negotiate, or/and—considering the scenario of social anomie—a social pact for a better world with the commitment of entrepreneurs and governments).

The premise of less working hours, and therefore a better work-life balance, addresses another challenge of the sector. In many countries, piece-work is the predominant wage form for temporary workers. Many are forced to work long hours. Others choose to do so, either because the rates of pay are low or because they want to earn as much as possible while work is available. This is particularly so in the case of workers who have migrated from the countryside or from overseas. A ten to 12-h day, for six days per week, is the norm. Even in the United Kingdom, research revealed that self-employed construction workers are paid per shift (rather than per hour) and the shift is normally 10 to 12 h per day for six days a week (International Labour Office, 2001; Mella & Savage, 2017). In sum, if the technologies under scrutiny in this chapter increase productivity and reduce the working time (without reducing wages), this will obviously have a beneficial impact for the workers.

The final point, sustainable production methods, is also welcome in a sector like construction which is known for uncertainty and periods of boom and bust, with serious consequences for the stability of employment (International Labour Office, 2001, 2015; Lawrence & Werna, 2009; Mella & Savage, 2017). If the introduction of smart grids and homes lead to sustainability, it may generate steadier employment.

Smart grids entail additional apparatuses, but also have a component of substitution—i.e. replacing traditional types of grid. Therefore, the net employment generation will depend on how many jobs are gained in the new grids versus those who are lost through substitution. Smart homes entail additional apparatuses, more than substitution. Therefore, their proliferation may entail the creation of new jobs. At the same time, the methods to be used in the manufacturing and installation (capital- or labour-intensive) also play an important role.

It is important to highlight that at the moment we are dealing with scenarios, without a firm knowledge of what will actually happen, as the future depends on many variables. It is not clear if the new technologies will lead to employment generation, longstanding employment or the contrary. Nevertheless, it is useful to identify scenarios and its possible consequences.

<u>Challenges</u>

Following the structure of sub-Sect. 2.1, the challenges will be analysed in turn, including: increase in inequality, increase in casualization of work, lack of social dialogue, job losses and lack of socialization.

There is already large inequality within the workforce in the build environment, which is pyramidal. New technologies require highly-skilled workers. This will naturally increase inequality within the industry, if the bottom of the pyramid remains with the type of semi- and unskilled work that it has now. If there is a fully-fledged increase in the skilling, inequality will diminish. A scenario is that the semi- and unskilled workers will lose their jobs—or be replaced—as they may not be able to catch up with the new knowledge necessary to work with the new technologies.

Casualization of work, which is already high in the construction industry (Mella & Savage, 2017; International Labour Office, 2015; Lawrence & Werna, 2009), may increase not necessarily because of smart grids and smart homes themselves, but due to the trends associated with them and explained in Sect. 2.1.

The present situation regarding social dialogue is somehow similar to casualization. According to the aforementioned authors, it is also a concern in the construction industry. The products of construction are fixed in space, and this would have given the opportunity to construction workers to meet more frequently and strengthen their positions—as opposed to workers who never meet each other. However, the high levels of casualization, informalization and large-scale sub-contracting chains have weakened the organization of the workers, although there resistance remains.

The tendency is that the components of smart grids and smart homes are prefabricated off-site and only assembled in-site, and in some cases with automation. This does not help social dialogue. This reinforces a suggestion made before in this chapter: workers may benefit from ICT to speed up exchange of information, online dialogue among them and decision-making. It may be more realistic to explore potential benefits of technological development rather than to fight them.

The next point, job losses, follows the analysis for urban workers in what regards the capital- versus labour-intensity. The installation of grids and the construction homes (smart or not) can be either capital- or labour-intensive (or somewhere in between). This will naturally impact on employment generation or losses. Considering the drive for digitalization, it is likely that the tendency will be towards capital-intensive methods. The example of 3 /4 D construction, made in Sect. 2.1.2 also applies here.

It is clear that technological innovations will create new jobs in the construction industry. But the rate between jobs creation and job losses is still uncertain. It is also clear that there is a tendency that the new jobs will migrate from the Global South to the North, and may also 'migrate' from senior workers to young ones. Again there is uncertainty in regard to the fate of those whose jobs have migrated away from them.

The possible effects of unemployment, already analysed for urban workers in general, also apply for construction workers in particular. This leads to lack of socialization, which can also happen with the dispersion of workers in different workplaces and lack of physical contact. The latter may be milder in the assembly of smart grids and homes because it still requires groups of workers working together. Nevertheless, it is important to monitor how the construction methods evolve and to be aware of possible consequences.

4 Conclusion

The Chapter analysed key technological developments which have clear potential to improve the life in cities and towns. Usually, the literature focus on the benefit of technological development to people as residents/users. But at the same time, they are workers. It is important to understand the impact on the first because one needs to have a decent income to live decently in a city or town. At the same time, the workers are the ones who produce and install the new equipment, therefore they need to be skilled and in good conditions to do so.

The actual impact on the workforce is still a subject of inquiry, as it will depend on many variables, noted along the Chapter. Still, it is possible to detect some trends, and at the same time scenario-building is important to be prepared for the consequences.

As in the case of any radical technological innovation, workers need to be trained or re-trained. In advanced economies, some level of training in such innovations are already taught in TVET (technical and vocational training) institutions for future construction workers or in retraining courses for current workers. How advanced it is, and how relevant to the needs of the labour market, is a different question, and there is large variability. In less developed countries such skills are nascent as is related training.

The Chapter analyzed the impact on the quantity and quality of employment. Those in the construction industry who are hired to work in the production and installation of the new equipment will naturally benefit. Many workers in other trades have also been able to benefit from the innovations. At the same time, dispersion have a tool on casualization, social dialogue and socialization. The recommendation is that workers should explore how ICT could speed up exchange of information, online dialogue among them and decision-making. It is more realistic to explore potential benefits of technological development rather than to fight them.

Unemployment should be carefully monitored, for all the reasons explained in the Chapter. It may happen due to substitution of humans by machines, due to 'migration' of jobs from one region to another, due to difficulties of part of the workforce to retrain (bearing in mind that the little trained 'bottom of the pyramid' in construction is vast). In a scenario of unemployment, solutions related to social protection have to be in place. Social pacts may also be important to stimulate solidarity (Sect. 2.2.1.).

Dispersion and disaggregation have impacts not only to work practices but also to the concept of the 'urban' in general. The urban economy, heavily based on the benefits of agglomeration, has been an important driver for the way cities and towns have developed. This could change in the future, leading to more dispersed settlements.

One way or another, the sweeping changes that the new technologies are about to bring must be embedded in an environment-conscious approach. Economic growth based on the ever-expanding production of new goods is not infinite.

References

Amsterdam Smart City. (2019). *Amsterdam Smart City*. Retrieved from https://amsterdamsmartcity.com.

Anthopoulos, L., & Fitsilis, P. (2010, July). From digital to ubiquitous cities: Defining a common architecture for urban development. In *2010 Sixth International Conference on Intelligent Environments* (pp. 301–306). IEEE.

Asres, M. W., Ardito, L., & Patti, E. (2021). Computational cost analysis and data-driven predictive modeling of cloud-based online NILM algorithm. *IEEE Transactions on Cloud Computing* 1–1.

Bahmanyar, A., Jamali, S., Estebsari, A., & Bompard, E. (2017). A comparison framework for distribution system outage and fault location methods. *Electric Power Systems Research, 145*, 19–34.

Bahmanyar, A., Estebsari, A., Pons, E., Patti, E., Jamali, S., Bompard, E. & Acquaviva, A. (2016). Fast fault location for fast restoration of smart electrical distribution grids. In *2016 IEEE International Smart Cities Conference (ISC2)* (pp. 1–6). IEEE.

Barbierato, L., Estebsari, A., Pons, E., Pau, M., Salassa, F., Ghirardi, M., & Patti, E. (2019). A distributed IoT infrastructure to test and deploy real-time demand response in smart grids. *IEEE Internet of Things Journal, 6*(1), 1136–1146.

Batic-Mujanovic, O., Poric, S., Pranjic, N., Ramic, E., Alibasic, E., & Karic, E. (2017). Influence of unemployment on mental health of the working age population. *Mater Sociomedica, 29*(2), 92–96.

Bompard, E., et al. (2016). A multi-site real-time co-simulation platform for the testing of control strategies of distributed storage and V2G in distribution networks. In *2016 18th European Conference on Power Electronics and Applications (EPE'16 ECCE Europe)*. IEEE.

Borlase, S. (Ed.) (2017). *Smart grids: Infrastructure, technology, and solutions*. CRC Press.

Bottaccioli, L., Estebsari, A., Pons, E., Bompard, E., Macii, E., Patti, E., & Acquaviva, A. (2017). A flexible distributed infrastructure for real-time cosimulations in smart grids. *IEEE Transactions on Industrial Informatics, 13*(6), 3265–3274.

Buckley, M. (2018). Labour and the city: Some notes across theory and research. *Geography Compass, 12*(3), 81–94.

Cambridge.org. (2019). *Cambridge Dictionary | English Dictionary, Translations & Thesaurus*. Retrieved from https://dictionary.cambridge.org.

CEN-CENELEC-ETSI Smart Grid Coordination Group Smart Grid Reference Architecture. (2012). Retrieved January 10, 2020, from https://ec.europa.eu/energy/sites/ener/files/documents/xpert_group1_reference_architecture.pdf.

Chiemezie, O. N., Gloria, A. C., & Gordon, C. O. (2013, November). Towards achievement of adequate consumer integration: The smart grid approach. In *2013 IEEE International Conference on Emerging & Sustainable Technologies for Power & ICT in a Developing Society (NIGERCON)* (pp. 314–318). IEEE.

Conejo, A. J., Morales, J. M., & Baringo, L. (2010). Real-time demand response model. *IEEE Transactions on Smart Grid, 1*(3), 236–242.

Costa H. A. (2021) Social dialogue and the future of work. In W Leal Filho, AM Azul, L Brandli, A Lange Salvia, T Wall (Eds). *Decent work and economic growth. Encyclopedia of the UN sustainable development goals.*

dictionary.cambridge.org. (n.d.). *smartness*. Retrieved from https://dictionary.cambridge.org/dictionary/english/smartness.

Dolber, B., Rodino-Colocino, M., Kumanyika, C., & Wolfson, T. (2021). *The gig economy: Workers and media in the age of convergence*. Routledge.

Dyson, M. E. H., Borgeson, S. D., Tabone, M. D., & Callaway, D. S. (2014). Using smart meter data to estimate demand response potential, with application to solar energy integration. *Energy Policy, 73*, 607–619. Retrieved April 5, 2020, from https://www.sciencedirect.com/science/article/pii/S0301421514003681.

Van Bastelaer, B. (1998). Digital cities and transferability of results. In *4th EDC Conference on Digital Cities, Salzburg, October 29–30, 1998* (pp. 61–70).
Van Empel, C., Werna, E. (2008). Labour oriented participation in municipalities: How decentralized social dialogue can benefit the urban economy and its sectors. *Sectorial Policies Department Working Paper 280.* ILO: Geneva.
Estebsari, A. (2016). An improved fault location method for distribution networks exploiting emerging LV smart meters. In: *IEEE Workshop on Environmental, Energy, and Structural Monitoring Systems (EESMS).* IEEE.
Estebsari, A., Orlando, M., Pons, E., Acquaviva, A., & Patti, E. (2018, September). A novel internet-of-things infrastructure to support self-healing distribution systems. In *2018 International Conference on Smart Energy Systems and Technologies (SEST)* (pp. 1–6). IEEE.
Estebsari, A., Barbierato, L., Bahmanyar, A., Bottaccioli, L., Macii, E., & Patti, E. (2019). A SGAM-based test platform to develop a scheme for wide area measurement-free monitoring of smart grids under high PV penetration. *Energies, 12*(8), 1417.
Estebsari, A., & Rajabi, R. (2020). Single residential load forecasting using deep learning and image encoding techniques. *Electronics, 9*(1), 68.
Eugster, P. T., Felber, P. A., Guerraoui, R., & Kermarrec, A. M. (2003). The many faces of publish/subscribe. *ACM Computing Surveys (CSUR), 35*(2), 114–131.
Falahati, B., Fu, Y., & Wu, L. (2012). Reliability assessment of smart grid considering direct cyber-power interdependencies. *IEEE Transactions on Smart Grid, 3*(3), 1515–1524.
Felstead, A., & Henseke, G. (2017). Assessing the growth of remote working and its consequences for effort, well-being and work-life balance. *New Technology Work and Employment, 32*(1) , 11:28.
Ghatikar, G. (2016). *Internet of things and data analytics handbook* (pp. 495–512).
Global Commission on the Future of Work, International Labour Organization. (2019). *Work for a brighter future. Report.* ILO: Geneva.
Goger, G., Reismann, W., & Breitwieser, K. (2019). *National and International Impact of Digitalisation on Workers in Construction. Scientific Study for BWI (Building and Wood Workers International).* BWI: Geneva.
Hajiabadi, M., Farhadi, M., Babaiyan, V., & Estebsari, A. (2020). Deep learning with loss ensembles for solar power prediction in smart cities. *Smart Cities, 3*(3), 842–852.
Hughes, IntelliGrid Architecture Concepts and IEC61850. (2005/2006). IEEE/PES Transmission and Distribution Conference and Exhibition, Dallas, TX, USA, 2006, pp. 401–404. https://doi.org/10.1109/TDC.2006.1668526.
International Labour Office. (2001). The construction industry in the twenty-first century: Its image, employment prospects and skill requirements. Study for Tripartite Meeting. Sectoral Activities Department. ILO: Geneva.
International Labour Office. (2015). Good practices and challenges in promoting decent work in construction and infrastructure projects. *Issues paper for discussion at the Global Dialogue Forum on Good Practices and Challenges in Promoting Decent Work in Construction and Infrastructure Projects,* 19–20 November. ILO: Geneva.
International Workplace Group. (2019). *The IWG global workspace survey welcome to generation flex – the employee power shift.* Zug: IWG.
Ishida, T. (2002). Digital city Kyoto. *Communications of the ACM, 45*(7), 76–81.
Järventausta, P., Partanen, J., & Kärkkainen, S. (2008, September). Interactive customer gateway for electricity distribution management, electricity market, and services for energy efficiency. In *Proceedings of the 8th Nordic Distribution and Asset Management Conference (NORDAC), Bergen.*
Khajenasiri, I., Estebsari, A., Verhelst, M., & Gielen, G. (2017). *A Review on Internet of Things Solutions for Intelligent Energy Control in Buildings for Smart City Applications. Energy Procedia* (Vol. 111, pp. 770–779). Retrieved September 11, 2020, from https://www.sciencedirect.com/science/article/pii/S1876610217302692.

Kou, G., Markham, P., Purcell, T., Reynolds, A., Valdez, A., Orndorff, R., Starling, B., & VanSant, P. (2020). Renewable generation monitoring platform and its applications. *IET Generation, Transmission & Distribution, 14*(24), 5804–5808.

Kumar, N. M., Chand, A. A., Malvoni, M., Prasad, K. A., Mamun, K. A., Islam, F. R., & Chopra, S. S. (2020). Distributed energy resources and the application of AI, IoT, and blockchain in smart grids. *Energies, 13*(21), 5739.

Lawrence, R., & Werna, E. (Eds.). (2009). *Labour conditions for construction: Building cities, decent work and the role of local authorities*. Wiley-Blackwell.

Van Lieshout, M. J. (2001). Configuring the digital city of Amsterdam. *New Media & Society, 3*(2), 131–156.

London City Hall. (2013). *"Smart London" vision*. Retrieved July 1, 2021, from https://www.london.gov.uk/press-releases-5658.

Mazza, A., Rogin, A., Zalzar, S., Estebsari, A., & Bompard, E. (2019, September). Creation of a computational framework for the European transmission grid with Power-to-Gas. In *2019 54th International Universities Power Engineering Conference (UPEC)* (pp. 1–6). IEEE.

Mella, A., & Savage, M. (2017). Construction sector employment in low income countries. *ICED (Infrastructure and Cities for Economic Development) Report, DfID (Department of International Development, UK)*. DfID.

Mohseni, M., Joorabian, M., & Lashkar Ara, A. (2021). Distribution system reconfiguration in presence of Internet of things. *IET Generation, Transmission & Distribution.*

NIST. (2012). *Smart Grid Interoperability Panel*. Retrieved from https://www.nist.gov/programs-projects/smart-grid-national-coordination/smart-grid-interoperability-panel-sgip.

NIST. (2016). *Standards*. Retrieved July 1, 2021, from https://www.nist.gov/standards.

Oh, S., Haberl, J. S., & Baltazar, J.-C. (2020). Analysis methods for characterizing energy saving opportunities from home automation devices using smart meter data. *Energy and Buildings, 216*, 109955.

Pau, M., Patti, E., Barbierato, L., Estebsari, A., Pons, E., Ponci, F., & Monti, A. (2018). A cloud-based smart metering infrastructure for distribution grid services and automation. *Sustainable Energy, Grids and Networks, 15*, 14–25.

Pau, M., Patti, E., Barbierato, L., Estebsari, A., Pons, E., Ponci, F., & Monti, A. (2019). Design and accuracy analysis of multilevel state estimation based on smart metering infrastructure. *IEEE Transactions on Instrumentation and Measurement, 68*(11), 4300–4312.

Prostejovsky, A. M., Brosinsky, C., Heussen, K., Westermann, D., Kreusel, J., & Marinelli, M. (2019). The future role of human operators in highly automated electric power systems. *Electric Power Systems Research, 175*, 105883.

Remote Year. (2019). 10 remote work statistics you should know about. Retrieved March 4, 2021, from https://www.remoteyear.com/blog/10-remote-work-statistics-you-should-know-about.

Ren, Y., Fan, D., Feng, Q., Wang, Z., Sun, B., & Yang, D. (2019). Agent-based restoration approach for reliability with load balancing on smart grids. *Applied Energy, 249*, 46–57.

Retrieved from https://www.smartnation.gov.sg.

Sako, M. (2021) From remote work to working from anywhere. *Communications of the ACM, 64*(4), 20–22.

Salazar-Xirinachs, J. M.; Nübler, I., & Kozul-Wright, R. (2014). *Transforming economies - making industrial policy work for growth, jobs and development*. ILO & UNCTAD: Geneva.

sfenvironment.org - Our Home. Our City. Our Planet. (2011). *Environment Now's Work*. Retrieved from https://sfenvironment.org/article/building-a-green-workforce-environment-now/our-work.

Siano, P. (2014). Demand response and smart grids—A survey. *Renewable and Sustainable Energy Reviews, 30*, 461–478.

Smart Grids Task Force, SFTF, European Commission. (2009).

Stevic, M., Estebsari, A., Vogel, S., Pons, E., Bompard, E., Masera, M., & Monti, A. (2017). Multi-site European framework for real-time co-simulation of power systems. *IET Generation, Transmission & Distribution, 11*(17), 4126–4135.

Stevic, M., Vogel, S., Grigull, M., Monti, A., Estebsari, A., Pons, E., Huang, T., & Bompard, E. (2016, October). Virtual integration of laboratories over long distance for real-time co-simulation of power systems. In *IECON 2016–42nd Annual Conference of the IEEE Industrial Electronics Society* (pp. 6717–6721). IEEE.

Upadhyay, D., & Sampalli, S. (2020). SCADA (Supervisory Control and Data Acquisition) systems: Vulnerability assessment and security recommendations. *Computers & Security, 89*, 101666.

WBCSD (World Business Council for Sustainable Development). (2020). *Complex disruptions to the Future of Work: Macro trends, realities and market insights that shape the future of work.* Research Report. WBCSD: Geneva.

Wells, J., & Prado, M. G. (2019). Protecting the wages of migrant construction workers, part three: What can be learned from systems of wage protection in China, EU, US and Latin America. *Engineers Against Poverty. Research Report.*

Werna, E. (Ed.) (2008). Labour in Urban Areas", special issue of Habitat International, *32*(2),137–282.

Werna, E., & Klink, J. (forthcoming). The builders of cities- prospects for synergy between labour and the built environment. In E Ofori (Ed.). *Elgar research companion on construction economics.* Edward Elgar Publishing.

World Economic Forum. (2018). The future of jobs. Research report. Centre for the New Economy and Society. WEF: Geneva.

www.jisc.go.jp. (n.d.). *JISC-Japanese Industrial Standards Committee.* Retrieved from https://www.jisc.go.jp/eng/.

www.sgcc.com.cn. (n.d.). Retrieved from http://www.sgcc.com.cn/html/sgcc_main_en/col2017112614/column_2017112614_1.shtml.

Yuson, A. (2020). Interview about the future of work. Retrieved March 8, 2021, from https://www.youtube.com/watch?v=qLgmta6rF-Q.

Chapter 3
Artificial Intelligence Robot Safety: A Conceptual Framework and Research Agenda Based on New Institutional Economics and Social Media

Rita Yi Man Li and M. James C. Crabbe

Abstract According to "Huang's law", Artificial intelligence (AI)-related hardware increases in power 4–10 times per year. AI can benefit various stages of real estate development, from planning and construction to occupation and demolition. However, Hong Kong's legal system is currently behind when it comes to technological abilities, while the field of AI safety in built environments is still in its infancy. Negligent design and production processes, irresponsible data management, questionable deployment, algorithm training, sensor design and/or manufacture, unforeseen consequences from multiple data inputs, and erroneous AI operation based on sensor or remote data can all lead to accidents. Yet, determining how legal rules should apply to liability for losses caused by AI systems takes time. Traditional product liability laws can apply for some systems, meaning that the manufacturer will bear responsibility for a malfunctioning part. That said, more complex cases will undoubtedly have to come before the courts to determine whether something unsafe should be the manufacturer's fault or the individual's fault, as well as who should receive the subsequent financial and/or non-financial compensation, etc. Since AI adoption has an inevitable relationship with safety concerns, this project intends to shed light on responsible AI development and usage, with a specific focus on AI safety laws, policies, and people's perceptions. We will conduct a systematic literature review via the PRISMA approach to study the academic perspectives of AI safety policies and laws and data-mining publicly available content on social media platforms such as Twitter, YouTube, and Reddit to study societal concerns about AI safety in built environments. We will then research court cases and laws related to AI safety in 61 jurisdictions, in addition to policies that have been implemented

R. Y. M. Li (✉)
Department of Economics and Finance, Hong Kong Shue Yan University, North Point, Hong Kong
e-mail: ymli@hksyu.edu

M. J. C. Crabbe
Wolfson College, Oxford University, Oxford, U.K, England
e-mail: james.crabbe@wolfson.ox.ac.uk; james.crabbe@beds.ac.uk

Institute of Biomedical and Environmental Science & Technology, School of Life Sciences, University of Bedfordshire, Park Square, U.K, England

School of Life Sciences, Shanxi University, Taiyuan, China

R. T. M. Li et al. (eds.), *Current State of Art in Artificial Intelligence and Ubiquitous Cities*, https://doi.org/10.1007/978-981-19-0737-1_3

globally. Two case studies on AI suppliers that sell AI hardware and software to users of built environment will also be included. Another two case studies will be conducted on built environment companies (a contractor and Hong Kong International Airport) that use AI safety tools. The results obtained from social media, court cases, legislation, and policies will be discussed with local and international experts via a workshop, then released to the public to provide the international community and Hong Kong with unique policy and legal orientations.

Keywords Artificial intelligence · Robot · New institutional economics · PRISMA · Law · Case studies · Hong Kong

1 Introduction

Based on expert predictions, the world's population will grow by 4 billion over the next 60 years. With an additional 1,200,000 to 1,500,000 residential units being needed annually, traditional construction methods will not support such demand. Coupled with high labour costs and a labour shortage, there is a growing trend towards adopting more innovative approaches. In light of this, new methods such as prefabricated buildings and construction being carried out by AI robots with object detection will become more commonplace (Baghdadi et al., 2020). In Hong Kong, we are already using AI to identify people not wearing masks in housing estates and shopping malls during the Covid-19 pandemic. Building surveying and property management professionals have begun to use drones to detect cracks in concrete and water leaks via object detection.

Even though AI provides considerable levels of convenience and is generally prone to making fewer errors than humans, the danger of AI errors remains significant. Although computer program designs and algorithms are generally accurate, training a program requires sufficiently large, accurate, and non-biased data, which is usually challenging to acquire in the real world. Once applied to real-world instances, an AI design exposed to biased, false, or otherwise corrupted data can modify its 'understanding' to simply accept the biased or inaccurate information as accurate, thereby functioning according to erroneous data (Giuffrida, 2019). Moreover, strong AI, such as neuromorphic artificial general intelligence, appears to be the most dangerous.

Given that most people's time is spent within built environments and safety is an essential aspect of responsible AI (as well as corporate social responsibility) and there is a general lack of related studies in this field, we plan on studying academia's perspectives on AI safety across the four stages of real estate development: planning, construction, occupation, and demolition.

Although Plato was famous for advocating the need to leave policy decisions to the experts, others have argued for the importance of public comments (Awad et al., 2020 forthcoming). We can gauge global views on AI safety in built environments by studying publicly available social media comments. This data can then be compared with 61 jurisdictions with AI safety-related laws, policies, and court cases to understand the deviation between public needs and presently available statutes and policies. The results will provide local and international communities with unique policy orientations and offer insights into responsible AI development in the future. Revealing the differences in legal and policy design worldwide will also allow us to understand the impact of socio-cultural factors on AI safety laws and policy development.

Since successful law and policy implementation requires satisfying local needs, two case studies of AI suppliers and another two case studies of AI users who engage in the planning, construction, and occupation stages, will help us better understand their expectations regarding AI policies and laws. Finally, experts will discuss our results before a final policy and law recommendation for Hong Kong.

The objectives of this research are fivefold:

1. To study academia's and the public's concerns on AI safety in built environments globally;
2. To review existing AI safety policies and regulations in built environments;
3. To reveal the impact of six dimensions of culture and socio-economic development on global AI safety regulations and policies;
4. To study the safety concerns and experiences of AI users and suppliers in Hong Kong's built environment;
5. To provide the international community and Hong Kong with unique policy and law orientations based on the results mentioned above.

2 Literature Review

2.1 Applications of AI in Built Environments

Traditional real estate development includes planning, construction, occupation, and demolition stages. AI can be applied in all these stages.

In the planning stage, AI is a potential saviour with much potential for engaging computer companies (Yigitcanlar et al., 2021). Semantic enrichment of building information models (BIM)—using AI methods to supplement such models with explicit information derived using algorithms trained to recognise and infer predefined sets of target concepts within patterns of building data—can remove the need for normalisation (Sacks et al., 2020). Similarly, a combination of an artificial bee colony and a neural network has been used for approximating the safety factor of retaining walls (Ghaleini et al., 2019).

During the construction stage, AI techniques such as latent class clustering analysis, hybrid models of case-based reasoning and the analytic hierarchy process, and the artificial neural network backpropagation approach have all been used to implement safety assessments in megaprojects (Ayhan & Tokdemir, 2019).

The application of metaheuristic optimisation algorithms has been developed in planar (2D) structure designs. Improved black hole and multiverse algorithms for the optimal design of steel trusses and enhanced fireworks algorithms for the discrete sizing of steel skeletal structures have been used to reduce computational complexity. Moreover, 'topology optimisation'—a type of AI application—has been used in precast structures to find the optimal location of the elements, reducing the cost of trussed structures (Baghdadi et al., 2020).

During the occupation stage, the Internet of Things (IoT) and AI have been used to deliver improved energy efficiency and lower operational costs in smart cities (Boje et al., 2020). For example, the Internet of Robotic Things (IoRT) uses deep learning for smart cities. Care-0-Bot 4 provides a cost-effective approach to operating in different environments and supports a wide range of applications. It can be deployed in museums, airports, homes, and offices for security purposes to help with daily chores, monitor the elderly, and provide healthcare services to patients. AndyVision—an inventory assistant—helps shops by calculating various customer needs; while Packbot is a self-organising robot used to identify the location of bombs and other hazardous materials (Liu et al., 2020).

AI algorithms and robots have been used throughout the built environment. Yet, research on AI safety precedents, legislation, policy, and public opinions regarding built environments is scarce. As there are substantially more people who use and are affected by AI-related products in the occupation stage (e.g., AI robots for Covid-19 disinfection and heavy-duty transportation in warehouses and factories), we speculate that there are more AI safety issues in the occupation stage. Thus, we hypothesise that:

H1: Regulations, policies, and the public's (including academia) perception/opinion concerning AI safety in various stages of real estate development are different. There is vastly more concern with AI safety issues in the occupation stage.

2.2 *Ethical and Responsible AI: Safety Liability*

> AI gives rise to errors of inference, of decision, and of action, which arise from the more or less independent operation of artefacts, for which no rational explanations are available, and which may be incapable of investigation, correction, and reparation (Clarke, 2019).

Technologies, including AI, are intrinsically neutral, and humans (both developers and users) determine the purposes for which AI is applied (Feijóo & Kwon, 2020). The starting point for examining 'computer-generated torts' is that AI is much safer than people in terms of the risk of injury (Perc et al., 2019). That said, accidents will always happen, and regulating AI in several ways already exists in terms of common law, statutory and regulatory obligations for organisations, and tort (Cuéllar, 2019). However, determining how legal rules should apply to liability for losses caused by AI systems takes time. Rather than adapting or creating a full compensatory system for AI harm, we should focus on identifying, minimising, and perhaps compensating for the predicted harm with regular compulsory re-evaluation.

As systems become more autonomous, determining liability when humans are harmed becomes complicated (Jamjoom et al., 2020). For example, where a self-driving car appears to be doing something unsafe and the driver overrides it—was it the individual's or the manufacturer's fault is the crux of the issue (Perc et al., 2019). An accident may also be caused by defective AI design, training, sensor design, unforeseen consequences from multiple data inputs, erroneous AI operation based on sensor or remote data, or external interference that could have been accidental or intentional by either one or more private actors, or on behalf of a foreign organisation or nation. Notably, sensors from multiple companies may have never been used together. In such instances, causation and responsibility may be impossible to determine. *Res ipsa loquitur* (the thing that speaks for itself) which could allow a court to conclude that a defendant must have been negligent since it is the only reasonable explanation for such a harmful act, cannot be applicable (Giuffrida, 2019).

The coexistence and interaction of humans and AI contributes to issues related to people's safety (Dodig & Çürüklü, 2012), with a vital concern being hacking. What makes an AI robot useful—its strength, ability to access and operate in challenging environments, expendability, etc.—could also be turned against people, either by criminals or people causing mischief. This issue will become more important as robots become more networked and increasingly indispensable to everyday life, just as computers and smart phones are today (Lin et al., 2011). One might suggest that injuries caused by hacking could be avoided through better safety by design, encouraging product designers to consider safety concerns during development (Dodig & Çürüklü, 2012).

In short, broader liability can be observed, particularly in the entities involved—the producer, distributor, seller, and the telecommunications provider—when, for example, the accident was caused due to a lack of an Internet connection (Perc et al., 2019). While researchers raise the concern of different stakeholders' liability for AI safety, none have studied the relative liability of these parties. We hypothesise that:

H2: ***Res ipsa loquitur* is not applicable in AI safety in all built-up environments. Manufacturers, algorithm designers, and other stakeholders bear different levels of legal liability, including but not limited to financial consequences.**

2.3 AI Safety Legal Issue

2.3.1 AI Safety Legal Issue 1: Compensation for Injuries and Encouragement of Innovation

The idea of injury compensation provided by the wrongdoers can reduce negligence but often discourages putting innovative ideas into practice. A clear example comes from the National Childhood Vaccine Injury Act (Public Law 99–660) in the US, which bears the societal obligation to compensate those who suffer injuries associated with vaccination. The act created the Vaccine Injury Compensation Program (VICP) to collect taxes on each dose of vaccine to pay for compensation (Thompson et al., 2020). Similarly, New Zealand has a long-established no-fault compensation scheme to cover almost all injury costs, including medical procedure damages, regardless of who was at fault (Fosch-Villaronga & Millard, 2019). The experiences with autonomous cars suggest that we should encourage technological innovation while compensating those hurt by the new technology (Giuffrida, 2019). While most of the no-fault compensation research sheds light on the medical industry, research into AI in this regard is scarce. As such, we hypothesise that:

H3: Most countries apply no-fault compensation related to injuries caused by AI applications in built environments. AI safety policies and laws that compensate those hurt by AI also include policies that encourage AI innovation.

2.3.2 AI Safety Issue 2: Does AI Incur Product or Personal Liability?

Given that currently product liability is strict (that is, independent of fault), and human activity is measured according to the standard of a reasonable person, several legal scholars claim that it is necessary to treat a computer tortfeasor as a person rather than as a product, thus incentivising automation and further improving safety. This means that intelligent objects should be evaluated under a negligence standard rather than a strict liability standard and that liability for damage would be comparable to that of a reasonable person (Abbott, 2018). Additionally, suppose it is proven that computers are safer than people. In that case, this could set the basis for a new standard of care for humans, so that human acts would be assessed from the perspective of what a computer would have done; and how if they had used a computer, the human could have avoided the accident and the consequent harm (Perc et al., 2019).

In the European Union, the Product Liability Directive (85/374/EEC) states that when a defective product causes damage, it is recoverable under the Product Liability Directive which regulates strict liability tests. This means that where AI causes an accident due to malfunctioning software, the consumer can bring a claim under the Product Liability Directive against the producer of the software. However, when software is supplied over the Internet, potential defects do not fall within this directive. Thus, a specific order on the liability of suppliers of digital content is required. Article 2(1) of Directive 2001/95 on general product safety includes any product intended for consumer use or likely to be used by consumers "including in the context of providing a service" (Perc et al., 2019).

Although there is a discussion regarding damage caused by a defective product, it is unknown whether AI operating at any stage of real estate development—evaluated under either strict liability or negligence standards—is treated as a person or a product. We hypothesise that:

H4: Courts and policies treat AI that causes injury in built environments as a person rather than a product and evaluate intelligent object injury cases under negligence standards rather than strict liability standards.

2.4 Deviations Between the Implemented Policies and Regulations, and the Needs of the Public

A main line of research has always been whether existing policies meet the public's needs. Equally important are determining ways to improve them. For example, Lopez-Behar et al. (2019) present demand-focused policy interventions which need to be addressed, such as regulatory awareness/information and fiscal/financial measures. Financial/fiscal policy measures/aid are required to incentivise electric vehicle owners and building owners (e.g., encouraging retrofitting). Regulatory policy measures should address the rights and obligations of the stakeholders and mandatorily require the installation of charging stations in new multi-unit residential buildings. Im and Park (2020) revealed that ten crane safety Acts suffer from duplication, inadequacy, and conflict problems. They proposed avoiding using ambiguous expressions, strengthening relaxed specification requirements, and harmonising safety standards across related government authorities to ensure consistency of standards.

Despite safety being identified as one of the main challenges in AI (Dwivedi et al., 2019), there is no research into whether existing safety policies address the needs of AI users, suppliers, and sellers in built environments. Based on the literature as mentioned earlier, we hypothesise that:

H5: AI safety policies fail to address the needs/concerns of AI users, suppliers, and sellers in built environments

2.5 New Institutional Economics, Formal and Informal Institutions

Under the lens of new institutional economics, formal (law and policies) and informal (culture and norms) institutions affect people's behaviours. Regarding formal institutions, Tu et al. (2020) examined the market reaction to firms in heavy-polluting industries and the effects of external legal, institutional quality and internal environmental disclosure on firm value around the passing of the Environmental Protection Tax Law in China. Li and Poon (2013) reviewed construction accident compensation in the United Kingdom and Hong Kong, while (Li & Chau, 2016) studied the impact of superstition (culture) on residential prices in Hong Kong.

Gu et al. (2020) suggested that G7 countries develop their high-tech industry and proposed policies that encourage innovation through research and development—an increasingly important engine for technological progress. It is well known that government regulations promote improvements that incorporate environmental standards (Sun et al., 2020).

2.5.1 Global Law and Policy Analysis and Review

Comparative laws promote international harmonisation and unification. They reveal the common core of all law, increase understanding of foreign rules, concepts, and approaches, and facilitate communication among foreign lawyers (Junker, 2014). Lavelanet et al. (2020) reviewed 158 countries' laws, policies, standards and guidelines, and legal judgments related to barriers to abortion. Of the countries studied, 107 provided information about health care providers who can conduct induced abortion services. Michalek et al. (2019) found that 69 countries out of 193 in the United Nations have regulations concerning the permissible level of heavy metals in cosmetics. While legal rules cover most high-income countries, none of the low-income countries has related laws. Using panel data of the legislative activity in 133 countries from 1999 to 2016, Eskander and Fankhauser (2020) found that parliamentary acts are driven by countries with a strong rule of law.

Kanter et al. (2020) assembled a global nitrogen policies database by reviewing 2,726 policies across 186 countries' legislatures and government agencies. They found that sink-centred policies predominantly focused on water. Most of the literature has not investigated the global landscape of AI safety-related policies and regulations. Thus, this research project aims to fill this gap and hypothesises that:

H6: AI safety laws and policies predominately cover most high-income countries' built environments, with predominant coverage on one (or more) area(s).

H7: Countries with a stronger rule of law are more likely to have laws and policies related to AI safety.

2.5.2 Cultural Impact on AI Safety Policies and Regulations

According to the Turing Institute, AI ethics is a set of values, principles, and techniques that utilise widely accepted standards of right and wrong to guide AI's moral development and use. For example, a recent study by MIT, in which an autonomous vehicle was out of control and needed to choose between crashing into an old or young man and a dog or a human being, looked into moral machine research related to ethical concerns of AI safety. The results, which took into account 40 million participants' decisions across 233 countries, showed that people from collectivistic cultures, which emphasise respect for the elderly, have a weaker preference to save younger people (Awad et al., 2018). At the same time, most respondents exhibited a moral choice for autonomous vehicles sacrificing their passengers to save more pedestrians (Bonnefon et al., 2016).

Apart from the collectivism mentioned above, societies mainly differ in five core cultural values: masculinity, power distance, long-term orientation, uncertainty avoidance, and indulgence (Zotzmann et al., 2019). All of these may have an impact on the laws and policies that are implemented.

Likewise, legal-economic theorists argue that common law and civil law develop differently. For example, a study of the rigidity of 86 countries' labour regulations from 1950 to 2004 evidenced the negative impact of individualism on labour regulations, with rigidity being enhanced by the legal system. Individualism has no effect on the rigidity of labour market regulations in civil law countries. While individualists see the government as a possible barrier to achievement, collectivists feel that a benevolent government creates stability through active intervention. Consequently, collectivists will have a more favourable view of government labour market regulations (Ang & Fredriksson, 2018).

One of the most attractive features of interpretive policy analysis is to provide a detailed and nuanced account of how policy-makers' actions shape and are shaped by broader cultural forces (Rogan, 2018). Although culture does affect labour regulations, there is a lack of research into whether cultural factors influence AI regulation and policy implementation. As such, we hypothesise that:

H8: Six dimensions of national culture and legal systems impact the likelihood of regulating AI safety via laws and policies.

2.6 *Social Media and Policy Making*

As of January 2018, roughly 4.021 billion people worldwide used the Internet. Of these, 3.196 billion people used social media. Such activities generate massive amounts of text data. Despite most data being unstructured (Curiskis et al., 2020), with people increasingly using various social media tools (e.g., YouTube, Reddit, Twitter, etc.) to express themselves about different issues (e.g., delivery of government services), their positive or negative comments (Belkahla et al., 2019; Yao et al., 2021) can complement, enrich, or even replace conventional data. Hence, these comments may hold great value and incur lower costs (Gal-Tzur et al., 2014).

Thus, extracting people's comments comprehensively from social media has become important for policy makers, and they may consider taking into account this information when making decisions (Belkahla et al., 2019). Social media opinion mining is also better than surveys and interviews because the latter often have meagre response rates.

While previous research has found that social media provides valuable information when formulating transport policy—the outcomes were highly relevant to a wide range of stakeholders in the transport industry (Gal-Tzur et al., 2014)—its usefulness in studying people's perceptions of AI safety in terms of built environment policies and regulations are unknown. Thus, we hypothesise that:

H9: Social media content provides valuable information regarding AI safety issues in the built environment for policy makers and legal drafters.

3 Research Method

The data will be collected from four primary sources: (1) academia's perspectives on AI safety in built environments will be collected by a systematic literature review; (2) the public's viewpoints, demands, and expectations regarding AI safety in built environments will be collected by studying the comments expressed by the public on social media; (3) global AI safety policies and regulations will provide a picture of the current institutional framework, and (4) case studies will provide insights regarding the needs and concerns of Hong Kong's AI users and suppliers in terms of built environments. The four sources will then be integrated to show the supply and demand of AI safety policies and regulations, thereby informing policy makers on which current policies do not meet public expectations. The results will also provide the international community and Hong Kong with unique policy orientations, which can be developed globally as well as locally.

3.1 Systematic Literature Review

Journal articles indexed in Web of Science, Science Direct, ASCE, Hein Online, West Law, and IEEE between 2001 and 2021 will be analysed utilising the systemic review approach. Systematic literature review offers a more accurate and broader level of understanding than a conventional literature review.

We will follow the Preferred Reporting Items for Systematic Reviews and Meta-Analyses (PRISMA)—an evidence-based approach for conducting systematic reviews that involves identification, screening, eligibility, and included stages, and which has been used for articles in many leading journals such as The Lancet and JAMA. It was also used to systematically review the AI technologies' challenges and opportunities in the construction industry (Regona et al., 2022). It involves the following steps (Fig. 1):

1. Identification: We will use 'AI safety' and 'artificial intelligence safety' for the keyword search in the aforementioned electronic databases.
2. Screening: We will only include articles' titles and abstracts that are relevant to real estate development (planning, construction, occupation, and demolition stages) and smart cities and will remove duplicate and irrelevant ones.

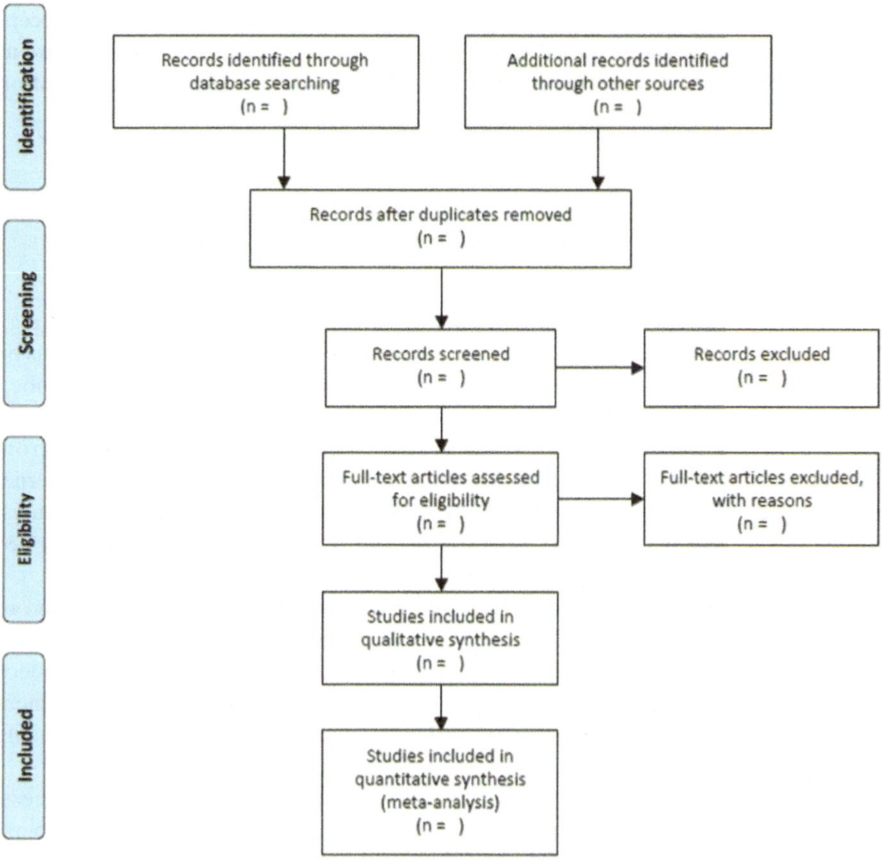

Fig. 1 PRISMA systematic literature review (Moher et al., 2009)

3. Eligibility: Based on no. 2, we will review the articles which look relevant and identify suggestions regarding the concerns, legislation (statutes), court cases, and policies related to AI safety.
4. Included: Only relevant articles will be used for content analysis (Fig. 1).
5. Content analysis: We will group the results of AI safety concerns according to (1) stages of real estate development, namely, planning, construction, occupation, demolition, and urban renewal; and (2) locations.

We can then compare people's main concerns over AI safety across the different stages of real estate development. Similar categories of concerns will be grouped in the form of content analysis. The content will also be studied through tokenisation, a natural language processing approach, as per Li et al. (2019). Phrases that occur most frequently will be ranked and reviewed. This will fulfil:

H1: Regulations, policies, and the public's (including academia) perception/opinion concerning AI safety in various stages of real estate development are different. There is vastly more concern with AI safety issues in the occupation stage

3.2 Social Media Text and Data Mining

To study the AI safety concerns of ordinary individuals who may be users of AI (e.g., smart home products or construction sector tools), we will explore public opinions by collecting people's comments on social media (e.g., Reddit, Twitter, and YouTube). Following Park et al. (2018), we will use PRAW (Python Reddit API Wrapper) to obtain discussion content on 'AI safety' and 'artificial intelligence safety' from the Reddit community r/safety. This public community allows anyone with Internet access to view the content. The two keywords, as mentioned earlier with hashtags, will be used to collect Tweets via NodeXL. We will subsequently review all scrapped content and only retain that related to any stage of built environment AI safety.

Regarding YouTube content, we will follow Devendorf et al. (2020) and clear our Internet browser history and cache, log off all accounts, and use Google Chrome's Incognito mode to reduce bias from our computer. We will search for relevant videos using the "relevance" filter, which will be viewed manually to screen content pertinent to any stage of the built environment. Relevant content will be downloaded using youtube-transcript-API.

Only publicly available social media sites will be mined during the whole process, with the results presented in an aggregated format. The name, individual members' identity, and unique resource locator of each comment on the social media platform will not be disclosed. Social media sites with automated web crawlers restricted access, or blocked access by a robots.txt file will be excluded (Monkman et al., 2018). The exact wording of the people's opinions will be rewritten to protect individual identities from being disclosed.

As in stage 1, we will group the results of public opinions according to stages of real estate development and locations for those which can identify locations. Content analysis will be used to discover group content concerns over AI safety in built environments. In contrast, tokenisation will be used to determine the most frequently raised concerns. This will fulfil:

H1: Regulations, policies, and the public's (including academia) perception/opinion concerning AI safety in various stages of real estate development are different. There is vastly more concern with AI safety issues in the occupation stage.

H9: Social media content provides valuable information regarding AI safety issues for policy makers and legal drafters in the built environment.

3.3 Court Cases, Legislation, and Policy Study

Regulations and precedents can provide information regarding the legislators' concerns and disputes over AI safety issues. The proposed research will cover countries with publicly available court cases and regulation databases. Our preliminary research indicated the number of documents in these databases (Table 1). As of 15 July 2020, 61 out of 123 countries (a complete country list can be found at http://www.worldlii.org/countries.html) in the World Legal Information Institute have AI safety-related court cases, policies, and legislation. We will first review the title and abstract/introduction (if applicable). We will then remove those which are unrelated to built environments and analyse the relevant ones via content analysis. Special precedent will be given to case studies.

AI-related policy data will also be collected using the main search engines (i.e., Google and Yahoo) with 'AI safety policy' and the corresponding country name and 'artificial intelligence and the country name. We will go through the results manually to determine which ones are related to built environments.

After collecting policy and legislation data, we will then analyse them using the content analysis method. As most leading journals request open data, we will modify (Kanter et al., 2020) the results that we obtain about global environmental policies to create our database of AI safety international policies and regulations in each of the four stages of real estate development. Each stage will then be subcategorised into: (1) name of the policy; (2) financial incentives for AI safety development; (3) compensation for AI-related accidents; (4) board objectives related to AI safety; (5) data and methods, including reporting protocols; and (6) research and development, e.g., funding according to different stages of real estate development. A similar process will be repeated for the regulations. Based on the six categories mentioned above, we will be able to provide suggestions for AI safety policy criteria and give the international community a unique policy orientation. This will fulfil:

H3: Most countries apply no-fault compensation related to injuries caused by AI applications in built environments. AI safety policies and laws that compensate those hurt by AI also include policies that encourage AI innovation.

H4: Courts and policies treat AI that causes injury in built environments as a person rather than a product and evaluate intelligent object injury cases under negligence standards rather than strict liability standards.

H8: Six dimensions of national culture and legal systems impact the likelihood of regulating AI safety via laws and policies.

Zhao et al. (2019) analysed data based on cluster and content analysis. While we will analyse court cases via thematic analysis; we will include analysis on judges' views regarding (1) level of financial and non-financial liability borne by different stakeholders for injuries caused by AI, (2) whether the court adopts negligence/liability standards for AI-related injury cases, (3) whether the court perceives AI that causes

Table 1 Legal databases that will be used and the number of 'artificial intelligence' cases available as of 15 July 2020 (searching for phrases such as 'artificial intelligence and injury' in the databases will filter out many relevant cases in these databases)

Legal databases	Locations covered	Search results included	No. of documents
WorldLII World Legal Information Institute	89 databases from 61 jurisdictions via 14 legal information institutes	International decisions, law journals, law reform treaties, international law	5438
CommonLII Commonwealth Legal Information Institute	61 common law databases from 42 Commonwealth and common law jurisdictions via eight legal information institutes	Case law, legislation, legal materials	4490
AustLIISearch Australasian Legal Information Institute	24 databases from 18 jurisdictions	Case law, law review, law report	3214
BAILII (2000–2020) British and Irish Legal Information Institute	United Kingdom, England and Wales, Scotland, Northern Ireland, Jersey, Ireland, Europe	Case law, legislation	1876
HKLII	Hong Kong	Cases from Court of First Instance	5
NZLII New Zealand Legal Information Institute	New Zealand	Case law, legislation, law reform, treaties	305
SAFLII Southern African Legal Information Institute	Southern Africa	National gazettes, provincial gazettes, South Africa: municipal by-laws, South African national legislation, South African courts, tribunals, Labour courts, Court rolls	64
ULII Uganda Legal Information Institute	Uganda	Uganda statutory instruments (2), legislation as promulgated (1), judgment (4), government gazettes (3), consolidated legislation (1)	11
Asia AsianLII	63/23 databases from 17 Asian jurisdictions via 8 legal information institutes		276

(continued)

Table 1 (continued)

Legal databases	Locations covered	Search results included	No. of documents
CanLII	Canada (Federal) British Columbia Alberta, Saskatchewan, Manitoba, Ontario Quebec, New Brunswick, Nova Scotia, Prince Edward Island, Newfoundland and Labrador, Yukon Northwest Territories, Nunavut	Cases (1,533), legislation (82)	1615
CY LAW	Cyprus		4
IRLII Irish Legal Information Initiative	Ireland	Statutory instruments	2
Ittig Institute of Legal Information Theory and Technique			8
Jersey Legal Information Board	Jersey	Unreported judgments (5), JGLR article (3), contents pages (1), law as enacted (1)	10
LAWPHIL Philippine Laws and Jurisprudence Databank	Philippines	Search results enhanced by Google	27
LIIofIndia Legal Information Institute of India	India	Free, independent, and non-profit access to 159 databases of Indian law	111
MalawiLII Malawi Legal Information Institute		Judgment (5), Acts of Parliament (1)	6

injury in built environments as a person rather than a product, and (4) judges' interpretation of objectives of AI safety laws. This will fulfil:

H2: *Res ipsa loquitur* is not applicable in AI safety in all built environments. Manufacturers, algorithm designers, and other stakeholders bear different levels of legal liability, including but not limited to financial consequences.

H4: Courts and policies treat AI that causes injury in built environments as a person rather than a product and that intelligent object injury cases are evaluated under negligence standards rather than strict liability standards.

A world map will be generated using Tableau to show the geographic coverage of countries with AI safety regulations related to built environments. A scatter plot for countries with and without regulations will be generated, as per income, the rule of law, and adult literacy profiles, according to Michalek et al. (2019).

3.4 Piecewise Structural Equation Modelling

Structural Equation Modelling (SEM) has often been used to study the interrelationship between various factors. Zeng and Li (2021) utilised SEM to analyse the relationship between satisfaction, willingness to revisit, and experiential value of mountain health tourism sites. Li et al. (2019) utilised SEM to study factors that affect people's willingness to share construction safety knowledge via the IoT and web 2.0 (including all social media channels that allow people to read and write (Li & Pak, 2010; Li & Poon, 2011).

PiecewiseSEM, a type of structural equation modelling, is designed to analyse multivariate hypotheses and applies to categorical responses controlled by multiple factors. PiecewiseSEM paths represent causal relationships and account for correlations between various response variables (Robinson et al. 2020). Given the fact that policies/laws are implemented in one jurisdiction is categorical, some variables may have inter-relations. As such, we will follow Bowen et al.'s (2017) piecewiseSEM package to study the impact of six dimensions of culture and other socio-economic factors on the likelihood of implementing AI safety laws and policies. We will then report on the estimated path coefficient (Table 2).

The Tableau world map and structural equation modelling will fulfil the hypotheses:

H6: AI safety laws and policies predominately cover most high-income countries' built environments, with predominant coverage on one (or more) area(s).

H7: Countries with a stronger rule of law are more likely to have laws and policies related to AI safety.

H8: Six dimensions of national culture and legal systems impact the likelihood of regulating AI safety via laws and policies.

Table 2 Sources of data and hypothesis tests to be covered in the machine learning models

Data	Source of data (number of countries covered in the database)	Hypotheses to test for:
Binary dependent variable: **Model 1:** AI safety regulation (with regulation = 1, without = 0) **Model 2:** AI safety policy (with regulation = 1, without = 0)	World Legal Information Institute (61 out of 123 jurisdictions contain court cases, policies, and legislation related to AI safety as of 15 July 2020). Some policies may be found in Google	H6-H8
Independent variables		
Indices in Hofstede's model of six dimensions of culture: power distance (P), collectivism (C), masculinity (M), uncertainty avoidance (U), long-term orientation (L), and indulgence (I)	Hofstede Insights (117 countries) https://hi.hofstede-insights.com/national-culture	H8
Common law, civil law and mixed /other legal systems (the first variable common law = 1, mixed /other legal systems = 0, the second variable civil law = 1, mixed /other legal systems = 0) The interaction term between legal systems and the aforementioned cultural dimensions: common law*P, common law*C,…, common law*I; civil law*P,…, civil law*I etc	Central Intelligence Agency (256 countries) https://www.cia.gov/library/publications/the-world-factbook/fields/308.html	H8
Income ranking based on PPP adult literacy rate of the country	World Bank (2020) (217 countries for income and 264 countries for adult literacy) https://datacatalog.worldbank.org	H6
The rule of law index	World Justice Project (128 countries) https://worldjusticeproject.org/rule-of-law-index/global/2020	H7

3.5 *Case Studies*

A case study refers to the research of an individual or an event. It involves an in-depth record of an individual or a group. A case study enables a researcher to obtain detailed information from observed behaviours and events. Although a case study lacks a breadth of study, it can be so unique that it affects the findings' validity (Li & Poon, 2013). Much previous research has also adopted single or multiple case studies. For example, Sousa et al. (2016) conducted case studies with 25 participants from 15 countries to develop a procedural framework, information standards, and guidelines for forest management. The 15 case studies in Australia that study the possibility of co-development in social, economic and environmental aspects in construction projects and materials production companies aim to reduce construction waste.

Two case studies on AI suppliers that sell AI hardware and software to built environment users will be included to study their perceptions regarding AI safety policies in the industry. Similarly, we will also conduct two case studies, one focus on the main contractor that pioneers AI technology adoption during the planning and construction stages. The other may look at Hong Kong International Airport to study AI technology adoption during the occupation stage. Interviews with AI programmers, project managers, sales staff, and designers, as well as reviews of technical reports, websites, and on-site observations, will provide insights into future AI safety policy orientation on the local level.

4 Academic, Practical and Policy Contributions

AI tools and applications have become more prevalent in all stages of real estate development. Yet, research into policies and regulations in built environments is scarce. Our analysis provides a global review of the latest law and policy developments in the field and looks at their sociocultural impact. The research will also contribute practically, with industry practitioners, both users and suppliers, becoming more aware of AI safety policies and laws in built environments. It will also provide the international community and Hong Kong with unique policy and law orientation.

Policy relevancy to Hong Kong AI has been applied in various real estate development cycles (e.g., planning, construction, occupation, and demolition). It is expected that there will be even greater use in the coming decade. In addition to innovations such as object detection by unmanned vehicles to detect cracks in walls and autonomous trains (10 three-car trains in Hong Kong's Southern District have carried 170,000 passengers daily since 2016), AI robots are also being used in the homes of the elderly, schools, and shopping malls. Furthermore, smart home tools have started to be included in many new residential real estate projects in Hong Kong for a couple of years. Ultimately, smart home robots and similar technologies are being developed rapidly, and safety issues surrounding these innovations have become more pertinent.

Unfortunately, the legal framework has lagged behind, potentially giving rise to problems regarding responsible AI usage. In this project, we propose exploring 61 jurisdictions' overseas policy approaches to AI safety in the four stages mentioned above of real estate development. Relevant implemented policies may provide the international community and Hong Kong with unique policy orientations. Given that corresponding regulations often accompany policies to ensure compliance, we will also review the legislation and regulatory precedents in different jurisdictions. In general, a complete separation of law and policy from people's expectations, concerns, and morals may lead to non-compliance, leading to the tools becoming meaningless. Thus, modern policy makers often draw knowledge from outside the public domain, with academic experts often being involved in the process. Hence, academics' views on AI safety regulations, court cases, and policy will be solicited via the PRISMA 2009 approach, while the general public's views on AI safety policies and regulations will be explored via social media data and content analysis.

Acknowledgments This work described in this paper was fully supported by a grant from the Research Grants Council of the Hong Kong Special Administrative Region, China (Project No. UGC/IIDS15/E01/19).

References

Abbott, R. (2018). The reasonable computer: Disrupting the paradigm of tort liability. *George Washington Law Review, 86*, 1–45.

Ang, J. B., & Fredriksson, P. G. (2018). Culture, legal heritage and the regulation of labor. *Journal of Comparative Economics, 46*, 616–633.

Awad, E., Dsouza, S., Kim, R., Schulz, J., Henrich, J., Shariff, A., Bonnefon, F., Rahwan, I. (2018). The moral machine experiment. *Nature, 563*, 59–64.

Ayhan, B. U., & Tokdemir, O. B. (2019). Safety assessment in megaprojects using artificial intelligence. *Safety Science, 118*, 273–287.

Baghdadi A., Heristchian, M. & Kloft, H. (2020). Design of prefabricated wall-floor building systems using meta-heuristic optimization algorithms. *Automation in Construction, 114*, 103156.

Belkahla, D. O., Mellouli, S., & Trabelsi, Z. (2019). From citizens to government policy-makers: Social media data analysis. *Government Information Quarterly, 36*, 560–570.

Boje, C., Guerriero, A., Kubicki, S., & Rezgui, Y. (2020). Towards a semantic construction digital twin: Directions for future research. *Automation in Construction, 114*, 103179.

Bonnefon, J. F., Shariff, A., & Rahwan, I. (2016). The social dilemma of autonomous vehicles. *Science, 352*(6293), 1573–1576.

Clarke, R. (2019). Regulatory alternatives for AI. *Computer Law & Security Review, 35*, 398–409.

Cuéllar, M.-F. (2019). A common law for the age of artificial intelligence incremental adjudication, institutions, and relational non-arbitrariness. *Columbia Law Review, 119*, 1773–92.

Curiskis, S. A., Drake, B., Osborn, T. R., & Kennedy, P. J. (2020). An evaluation of document clustering and topic modelling in two online social networks: Twitter and Reddit. *Information Processing & Management, 57*, 102034.

Devendorf, A., Bender, A., & Rottenberg, J. (2020). Depression presentations, stigma, and mental health literacy: A critical review and YouTube content analysis. *Clinical Psychology Review, 78*, 101843.

Dodig, C. G., & Çürüklü, B. (2012). Robots: Ethical by design. *Ethics and Information Technology, 14*, 61–71.

Dwivedi, Y. K., Hughes, L., Ismagilova, E., Aarts, G., Coombs, C., Crick, T., et al. (2019). Artificial intelligence (AI): Multidisciplinary perspectives on emerging challenges, opportunities, and agenda for research, practice and policy. *International Journal of Information Management*, 101994.

Eskander, S. M. S. U., & Fankhauser, S. (2020). Reduction in greenhouse gas emissions from national climate legislation. *Nature Climate Change, 10*, 750–756.

Feijóo, C., & Kwon, Y. (2020). AI impacts on economy and society: Latest developments, open issues and new policy measures. *Telecommunications Policy, 44*, 101987.

Fosch-Villaronga, E., & Millard, C. (2019). Cloud robotics law and regulation: Challenges in the governance of complex and dynamic cyber–physical ecosystems. *Robotics and Autonomous Systems, 119*, 77–91.

Gal-Tzur, A., Grant-Muller, S. M., Kuflik, T., Minkov, E., Nocera, S., & Shoor, I. (2014). The potential of social media in delivering transport policy goals. *Transport Policy, 32*, 115–123.

Ghaleini, E. N., Koopialipoor, M., Momenzadeh, M., Sarafraz, M. E., Mohamad, E. T., & Gordan, B. (2019). A combination of artificial bee colony and neural network for approximating the safety factor of retaining walls. *Engineering with Computers, 35*, 647–658.

Giuffrida, I. (2019). Liability for AI decision-making: Some legal and ethical considerations. *Fordham Law Review, 88*, 439–456.

Gu, J., Umar, M., Soran, S., & Yue, X.-G. (2020). Exacerbating effect of energy prices on resource curse: Can research and development be a mitigating factor? *Resources Policy, 67*, 101689.

Jamjoom, A. A. B., Jamjoom, A. M. A., & Marcus, H. J. (2020). Exploring public opinion about liability and responsibility in surgical robotics. *Nature Machine Intelligence, 2*, 194–196.

Junker, K. W. (2014). A focus on comparison in comparative law. *Duquesne Law Review, 52*, 69–99.

Kanter, D. R., Chodos, O., Nordland, O., Rutigliano, M., & Winiwarter, W. (2020). Gaps and opportunities in nitrogen pollution policies around the world. *Nature Sustainability*.

Lavelanet, A. F., Johnson, B. R., & Ganatra, B. (2020). Global Abortion Policies Database: A descriptive analysis of the regulatory and policy environment related to abortion. *Best Practice & Research Clinical Obstetrics & Gynaecology, 62*, 25–35.

Li, R. Y. M., & Pak, D. H. A. (2010). Resistance and motivation to share sustainable development knowledge by web 2.0. *Journal of Information and Knowledge Management, 9*, 251–262.

Li, R. Y. M., & Poon, S. W. (2011). Using Web 2.0 to share knowledge of construction safety: The fable of economic animals. *Economic Affairs, 31*, 73–79.

Li, R. Y. M., & Poon, S. W. (2013). *Construction Safety*. Springer.

Li, R. Y. M., & Chau, K. W. (2016). *Econometric analyses of international housing markets*. Routledge.

Li, R. Y. M., Chau, K. W. & Zeng, F. F. (2019). Ranking of risks for the existing and new building works. *Sustainability, 11*, 2863.

Lin, P., Abney, K., & Bekey, G. (2011). Robot ethics: Mapping the issues for a mechanized world. *Artificial Intelligence, 175*, 942–949.

Liu, Y., Zhang, W., Pan, S., Li, Y., & Chen, Y. (2020). Analyzing the robotic behavior in a smart city with deep enforcement and imitation learning using IoRT. *Computer Communications, 150*, 346–356.

Lopez-Behar, D., Tran, M., Froese, T., Mayaud, J. R., Herrera, O. E., & Merida, W. (2019). Charging infrastructure for electric vehicles in Multi-Unit Residential Buildings: Mapping feedbacks and policy recommendations. *Energy Policy, 126*, 444–451.

Michalek, I. M., Benn, E. K. T., dos Santos, F. L. C., Gordon, S., Wen, C., & Liu, B. (2019). A systematic review of global legal regulations on the permissible level of heavy metals in cosmetics with particular emphasis on skin lightening products. *Environmental Research, 170*, 187–193.

Moher, D., Liberati, A., Tetzlaff, J., Altman, D.G., & The, P. G. (2009). Preferred reporting items for systematic reviews and meta-analyses: The PRISMA statement. *PLOS Medicine, 6*, e1000097.

Monkman, G. G., Kaiser, M. J., & Hyder, K. (2018).. Text and data mining of social media to map wildlife recreation activity. *Biological Conservation, 228*, 89–99.

Park, A., Conway, M., & Chen, A. T. (2018). Examining thematic similarity, difference, and membership in three online mental health communities from reddit: A text mining and visualization approach. *Computers in Human Behavior, 78*, 98–112.

Perc, M., Ozer, M., & Hojnik, J. (2019). Social and juristic challenges of artificial intelligence. *Palgrave Communications, 5*, 61.

Regona, M., Yigitcanlar, T., Xia, B., Li, R. Y. M. (2022). Opportunities and adoption challenges of AI in the construction industry: a PRISMA review. *Journal of Open Innovation: Technology, Market, and Complexity, 8*, 45.

Robinson, B. J. O., Barnes, D. K. A., & Morley, S. A. (2020). Disturbance, dispersal and marine assemblage structure: A case study from the nearshore Southern Ocean. *Marine Environmental Research, 160*, 105025.

Sacks, R., Girolami, M., & Brilakis, I. (2020). Building information modelling, artificial intelligence and construction tech. *Developments in the Built Environment*, 100011.

Sousa, C. D., Soares, A. L., & Pereira, C. S. (2016). Collaborative conceptualisation processes in the development of lightweight ontologies. *VINE Journal of Information and Knowledge Management Systems, 46*, 175–193.

Sun, S., Li, T., Ma, H., Li, R. Y. M., Gouliamos, K., Zheng, J., Han, Y., Manta, O., Comite, U., Barros, T., Duarte, N., & Yue, X.-G. (2020). Does Employee Quality Affect Corporate Social Responsibility? Evidence from China. *Sustainability, 12*, 2692.

Thompson, K. M., Orenstein, W. A., & Hinman, A. R. (2020). Performance of the United States vaccine injury compensation program (VICP): 1988–2019. *Vaccine, 38*, 2136–2143.

Tu, W.-J., Yue, X.-G., Liu, W., & Crabbe, M. J. C. (2020). Valuation impacts of environmental protection taxes and regulatory costs in heavy-polluting industries. *International Journal of Environmental Research and Public Health, 17*, 2070.

World Bank. (2020). GDP Ranking, PPP Based. Retrieved from https://datacatalog.worldbank.org/dataset/gdp-ranking-ppp-based.

Yao, Q., Li, R. Y. M., Song, L., & Crabbe, M. J. C. (2021). Construction safety knowledge sharing on Twitter: A social network analysis. *Safety Science, 143*, 105411.

Yigitcanlar, T., Corchado, J. M., Mehmood, R., Li, R. Y. M., Mossberger, K., & Desouza, K. (2021). Responsible Urban innovation with local government artificial intelligence (AI): A conceptual framework and research agenda. *Journal of Open Innovation: Technology, Market, and Complexity, 7*, 71.

Zeng, L., & Li, Y. M. R. (2021). Tourist satisfaction, willingness to revisit and recommend, and mountain kangyang tourism spots sustainability: A structural equation modelling approach. *Sustainability, 13*, 10620.

Zhao, X., Zuo, J., Wu, G., & Huang, C. (2019). A bibliometric review of green building research 2000–2016. *Architectural Science Review, 62*, 74–88.

Zotzmann, Y., van der Linden, D., & Wyrwa, K. (2019). The relation between country differences, cultural values, personality dimensions, and error orientation: An approach across three continents – Asia, Europe, and North America. *Safety Science, 120*, 185–193.

Chapter 4
The Impact of Artificial Intelligence Educational Robots in the Field of Education: a PRISMA Review

Xuanzhang Mou and Rita Yi Man Li

Abstract Education is an indispensable part of everyone's life. It lays the ground for the future generation and is a prerequisite for an independent life. It also determines the future of a country and a nation, and affects a country's development. With the advancement of technology, the world has tremendous changes. People have changed their lifestyles and use new technologies for connecting, interacting, reading, and obtaining information. Hence education needs to adapt to the new era and changes in social customs. This research throws light on the impact of artificial intelligence robots on the education sector. We initially found one hundred thirteen articles published between 1999 and 2019 and only 18 articles meeting the criteria. We then categorise these studies into the following topics: (1) the impact of artificial intelligence educational robots on the creativity and motivation of students' learning, and (2) the impact of artificial intelligence robots in education. This chapter comprises of six parts: the first part is the introduction, the second part summarises the purpose and research questions of this chapter, the third part is a literature review of educational robots concluding that robotics has a unique role in education, the fourth part discusses the research methods and survey results of this study, the fifth part provides a summary and discovery, and the last part is the conclusion.

Keywords Artificial intelligence · Educational robots · K-12 education · STEM education · Systematic literature review

1 Introduction

With the advancement of technology, robots have become part of our daily lives. For example, we see micro drones hover over us, Rosie and Roomba complete daily housework, wearable robotics and robots are used on construction sites to aid workers

X. Mou
Department of Economics and Finance, Hong Kong Shue Yan University, Hong Kong, China

R. Y. M. Li (✉)
Sustainable Real Estate Research Center, Hong Kong Shue Yan University, Hong Kong, China
e-mail: ymli@hksyu.edu

R. T. M. Li et al. (eds.), *Current State of Art in Artificial Intelligence and Ubiquitous Cities*, https://doi.org/10.1007/978-981-19-0737-1_4

in moving heavy materials and install large window panels on sites. Zenbo reminds older people in the elderly home to take medicine on time (Li, 2018; Li & Ng, 2018). In the education sector, integrating robots into the educational environment can cultivate students' interest in science, technology, engineering, and mathematics (STEM) topics while learning and understanding complex concepts more in-depth (Melchior et al., 2005).

Educational robots have been used in various fields, such as teaching materials (Lau et al., 1999; Wang, 2004), learning companions (Kory & Breazeal, 2014; Kory et al., 2013), and teaching assistants (Han & Kim, 2009). K-12 educational robots and robot competitions have become popular "teaching assistants" for implementing educational activities. They allow children to think critically and solve problems in a team environment. Hence, the number of studies regarding educational robots and their impact on the academic and skills of young learners is gradually increasing. The proportion of relevant documents published in 2015–2019 is considerable; approximately 70% of all papers were published during this time (Fig. 1).

Although some studies have suggested that the purpose of educational robots is to enhance students' interest and participation (Rubinstein et al., 2015), few investigate the effectiveness of educational robots about their roles in learning outcomes achievements and professional skills (e.g. communication, collaboration) enhancements. Although there is an increase in research on educational robots (Alimisis, 2013; Barker & Ansorge, 2007; Hendricks, & Ogletree, 2012; Menekse et al., 2017), few link theoretical foundations with reality.

Therefore, we aim to combine theoretical foundation with reality to study the effectiveness of artificial intelligence educational robots in education. A systematic review of relevant literature on educational robots in STEM education was conducted to study the impact of research educational robots on STEM education and student learning.

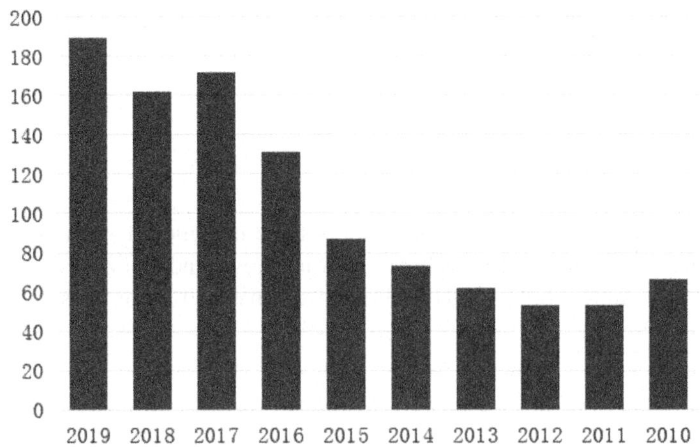

Fig. 1 Papers published and indexed in Web of Science and Google Scholar over the last ten years with keywords "AI" and "Education"

2 Research Purpose and Research Question

With an ever increase in applied artificial intelligence technology, we should not underestimate its impact on education. This research explores the impact of artificial intelligence robots on education, specifically how artificial intelligence educational robots in STEM enhance teaching interactiveness. The article reviews the importance of artificial intelligence robots in education.

The research questions are as follows:

(1) **Does the artificial intelligence educational robot support the learning of STEM concepts?**
(2) **Can artificial intelligence educational robots promote student learning?**

The research findings will help us understand how educational robots can be integrated into the traditional classroom and extracurricular STEM learning.

3 Literature Review

3.1 Educational Robotics

Robotics education refers to building, programming, and using robots to stimulate students' interest in learning while cultivating students' comprehensive abilities and critical thinking. In this process, students can master the basic structure and construction process of the robot and experience the fun of programming by controlling the robot. Therefore, robot education is entertaining (Bers & Urrea, 2000). The origin of educational robots can be traced back to the advent of the LOGO programming language in 1967, which developed educational robot teaching kid in K-12 STEM education. Educational robotics at that time was to cultivate children's skills and interest in computer programming.

In the 1980s, a pioneering report showed that children could solve problems by coding the "turtle" robot and learning the logo programming. This idea is based on educational robots' unique characteristics, which allow students to engage in computer science programming and non-coding tasks early (Papert, 1993).

Definition of selected keywords

LOGO—Computer language popular in elementary education and is utilised to teach mathematical concepts.

Robotics Education—The field of teaching robotics applications specifically within an engineering framework.

Educational Robotics—The use of robots to expand knowledge in various education areas.

Turtle—Term used to describe either an on-screen or hands-on robot commonly applied to Turtle Math or Turtle LOGO educational programmes in the computer. A drawing is usually attached to the actual robot to duplicate the design created on-screen by the computer-based robot.

3.2 Artificial Intelligence Robots in Education

Suppose a robot works with a teacher or colleague (robot) and is used for children's daily learning, including spelling and pronunciation, the learning plan and teaching methods can be adjusted according to different children's learning situations. This is an alternative education method (Timms, 2016; Fang et al., 2019). As early as 1980, after the development of Papert, some robots that used programming and building programmers to study the teaching of STEM concepts appeared (Knop, 2018). The early research on educational robots focused on computer programming because the founder Papert was one of the developers of the logo programming language.

Much research on educational robotics focuses on various concepts and skills related to STEM. Some of these studies have shown that educational robotics enhance students' critical thinking and problem-solving skills (Erdogan et al., 2013). Through the analysis of these studies, it is evident that educational robots can increase students' interest and participation in STEM (Kim et al., 2015).

Also, robotics can improve students' thinking skills (Khanlari & Kiaie, 2015). Robotics promotes students' innovative thinking and cultivates their problem-solving and teamwork skills. Therefore, artificial intelligence robots play a unique role in education.

3.3 Sub-Topic Discussion

After reviewing 18 articles, this study conducted classification and thematic analysis (Table 4). All articles were classified based on their research setting, design, and publication type. The articles were classified into the following two themes:

Theme 1: Artificial intelligence educational robots enhance the creativity and motivation of students in learning

Applied artificial intelligence on educational robots for student learning usually aims at improving students' learning ability and the efficiency in receiving new knowledge. Using artificial intelligence educational robots, students can actively participate in learning, they are encouraged to actively participate in the learning process (Sanchez Ruiz & Jamba, 2008).

With the help of artificial intelligence educational robots, the sense of accomplishment of specific learning tasks or the pleasure of comprehending new knowledge in a particular field will help engage students in learning (McGrath et al., 2008).

When students use artificial intelligence educational robot technology, they can learn by exploring new knowledge. These studies have demonstrated that: (a) autonomous learning through artificial intelligence educational robots allow students to understand some abstract concepts better (Krishnamoorthy & Kapila, 2016; Shankar et al., 2013) and (b) students' creativity and motivation can be improved through artificial intelligence robots (McKay et al., 2015).

As a model of this type of research, Williams and his colleagues evaluated a STEM extracurricular project which was implemented to explore the effectiveness of learning by participating various robotic activities. They believed that robots were a unique tool that could help students understand abstract mathematics concepts outside the traditional classroom environment. The researchers designed three interactive and team-based Lego activities and found that student participants improved their conceptual understanding of the content. Furthermore, students who participated in the activity were more interested and motivated to learn mathematics. It indicated that students could be more realistic and practical with mathematical applications through artificial intelligence education robots outside the classroom.

53 educational studies utilised robots to improve students' creativity and motivation in learning. These studies believe that social-cultural trends or creative motives increase students' motivation and interest in teaching, especially STEM and programming courses. The main idea of these studies is that using artificial intelligence robots in learning can raise students' interest in education and improve their learning ability. Moreover, we can stimulate students' creativity during the interaction with educational robots (Giannakos, Jaccheri & Proto, 2013; Hamner & Cross, 2013).

Theme 2: Artificial intelligence educational robots are helpful for education management

(1) **The role of artificial intelligence robots in education management**

One of the critical areas in education is education management. Artificial intelligence can significantly impact the education sector if it is applied to education management. For example, teachers can utilise artificial intelligence technology to comment and score students' homework and papers, and it will then be possible to give students one-to-one feedback to help them achieve excellent academic results, significantly improve the efficiency of education management (Rus et al., 2013). Artificial intelligence in education, especially distance and online education, will substantially enhance institutional efficiency and services (Sharma et al., 2019).

In today's learning environment, homework review has become the primary task of teachers, who should have to provide suggestions and grades for the students' assignments and check whether there is plagiarism. An artificial intelligence platform will help reduce the time required to complete this task, improving teachers' efficiency.

(2) **Instruction teaching**

Instruction teaching is to simulate some teaching instructions by applying artificial intelligence and other teaching tools, using intelligent technology different from reality to demonstrate the concepts and operational experiments in the textbook to learn through practice. This is a potent educational tool as the influx of artificial intelligence promotes the development of educational innovation systems and affects students' learning methods. Many studies emphasised the integration of artificial intelligence into machines or robots and creating more effective instructions for them to improve teaching strategies. The primary use of artificial intelligence in educational applications is integrating artificial intelligence into educational robots to assist educational robots in performing basic or advanced teaching tasks, such as teaching students pronunciation and literacy (Timms, 2016). Artificial intelligence and computer programs are combined to develop robots or online computers which is capable to understand dialogue and answer students' daily questions (Pokrivcakova, 2019).

The application of artificial intelligence in teaching was also highlighted. For example, the online education platform was a powerful tool for improving students' teaching efficiency and ability to absorb knowledge when teachers used AIWBES (Kahraman, et al., 2010). During this process, the teacher acts as the protagonist of the system's demonstration and systematically learns from the relevant organisation before teaching, then gives some instructions to the students in use. It can ensure that the technology used for education on the Internet is effective and systematic to improve the students' learning experience (Peredo et al., 2011).

An artificial intelligence education can bring personalised learning methods to students, customising them according to their needs. Artificial intelligence in tutoring and teaching systems solves the different challenges faced by one-on-one teachers and students, thereby improving the overall quality of teachers' work (Roll & Wylie, 2016). Hence, enhancing teachers' effectiveness and efficiency. The application of artificial intelligence in the education sector helps improve the effectiveness of instructions.

(3) **Learning**

Learning is a major channel for students to receive knowledge and is an integral part of education (UNESCO, 2019). Different types of artificial intelligence educational robots have been used to encourage students to learn, for example, the use of artificial intelligence for a specific application to improve the learning style of students and personalise the courses and content so that learners can engage in learning more happily (Wartman & Combs, 2018).

From another perspective, artificial intelligence in education eliminated the barriers to learning to a certain extent. For example, students could experience the education methods of different countries through the AIWBES platform, creating an atmosphere to learn without boundaries. Comprised of essential factors such as adaptive hypermedia, information filtering, monitoring, and collaborative learning, AIWBES was a platform that dedicates to learning. Learning through this platform could encourage cooperation and interaction between students (Kahraman et al., 2010).

Previous research discussed the relationship between AIWBES and learning quality improvement, emphasising that in AIWBES system could adjust according to customise instructions, determine and evaluate learners' learning content. At the same time, it also tracked the learning progress of learners and chose the right time for students to study (Peredo et al., 2011).

Although artificial intelligence can be used to encourage and cultivate academic integrity and help students revise their homework, it may also have some adverse effects on students' learning (Sutton, 2019). Crowe's team observed that artificial intelligence would promote the development of some online platforms that endanger academic integrity and might lead to students' plagiarism or dishonest behaviour (Crowe et al., 2017).

4 Research Method

4.1 Systematic Literature Review

A systematic literature review was conducted by searching, identifying and collating existing articles. We include articles indexed in ProQuest Central, IEEE Xplore, Google scholar, ASEE annual exposition and conference proceedings.

4.2 Theoretical Framework

The first step is to study the article comprehensively. Two aspects of artificial intelligence robots in education were addressed. One is to explore the impact of artificial intelligence robots on students' learning ability, and the other is to study the effects of artificial intelligence educational robots on education management. A theoretical framework was designed to understand the relationship (Fig. 2) better.

To better understand the meaning of the influencing factors, Table 1 introduces the main definitions in the study.

4.3 Analysis Methods and Data Sources

A systematic review is a statistical analysis of many individual studies designed to integrate their findings. In recent years, many searches have been conducted on literature. For example, Regona et al. (2022) reviewed the opportunities and adoption challenges of artificial intelligence in the Construction Industry by using Preferred

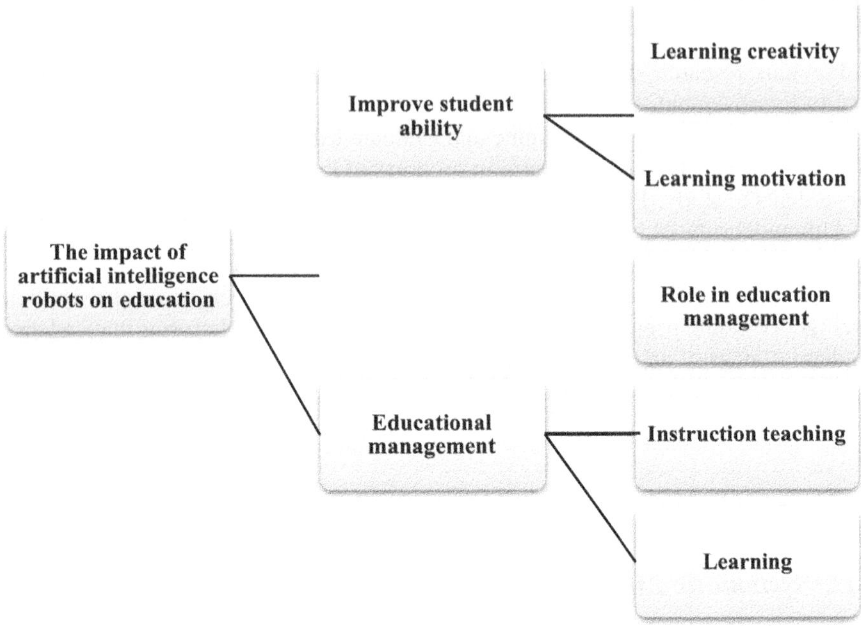

Fig. 2 The theoretical framework

Reporting Items for Systematic Reviews and Meta-Analyses (PRISMA) approach. This study included 113 studies. According to the exclusion criteria (Table 2), 95 articles were excluded based on the five exclusion principles (Table 2) and full-text review. Among them, there were reports of sample attributes that failed to meet the requirements (31), products from second or third sources (21), unrelated articles (18), irrelevant research (12), and incomplete or repeated (13) as shown in Fig. 3 (Regona et al., 2022).

According to the exclusion principle in Table 2, 18 studies were classified by subject (Table 3) and included in this research report (Table 4).

5 Research Results

(1) **Does the artificial intelligence educational robot support the learning of STEM concepts?**

Artificial intelligence in education is an extension of computer language and an innovative revolution in education. The artificial intelligence educational robot supports conceptual learning in the STEM classroom and operability-related learning outside the STEM classroom. Artificial intelligence robots have a robust data integration system and cloud computing capabilities, transforming

Table 1 Definitions of the terms used in this study

Areas of improvement	Reasons
• Learning creativity	Creativity is an innate ability that everyone has and can be improved through education and training. Professional creativity training can increase student creativity from 10 to 40%. The inner motivation of the individual to pursue success is the primary motivation for cultivating creativity
• Learning motivation	Learning motivation refers to the inner strength of a student for learning. A correct learning attitude is a prerequisite to ensure learning motivation and one of the critical elements to stimulate students' interest in education and improve learning motivation. When students become interested in learning, they will want to learn, love learning, and enjoy learning
• Role in education management	Educational management refers to the reasonable allocation of educational resources, including people, finances, time, space and information, and let them operate effectively to achieve organisational goals
• Instruction teaching	Guided teaching simulates some teaching guides by applying artificial intelligence and other teaching tools which are different from the reality to demonstrate concepts and conducting experiments in textbooks so students can learn from experience and practice
• Learning	Learning is the process of improving ability through external education or own experience. Specifically, it is the process of acquiring knowledge or skills through reading, listening, and observation. From a broad perspective, people gain experience during life

Table 2 Exclusion research criteria

Exclusion research criteria	
Article nature	The articles selected in this study are primarily expert interviews, or personal work and theoretical abstracts
Relevance	Articles that are not directly related to the content of this research
Sample attributes	Articles do not specify the age group or basic information required for this study
Secondary or tertiary source	Most of these articles are related studies that have undergone comparative studies by non-professionals or tried to predict
Repeat	Repetitive or incomplete with other articles

complex concepts in STEM into easy-to-understand knowledge and complex areas that require outdoor or laboratory practice into a virtual environment.

(2) **Can artificial intelligence educational robots promote student learning?**

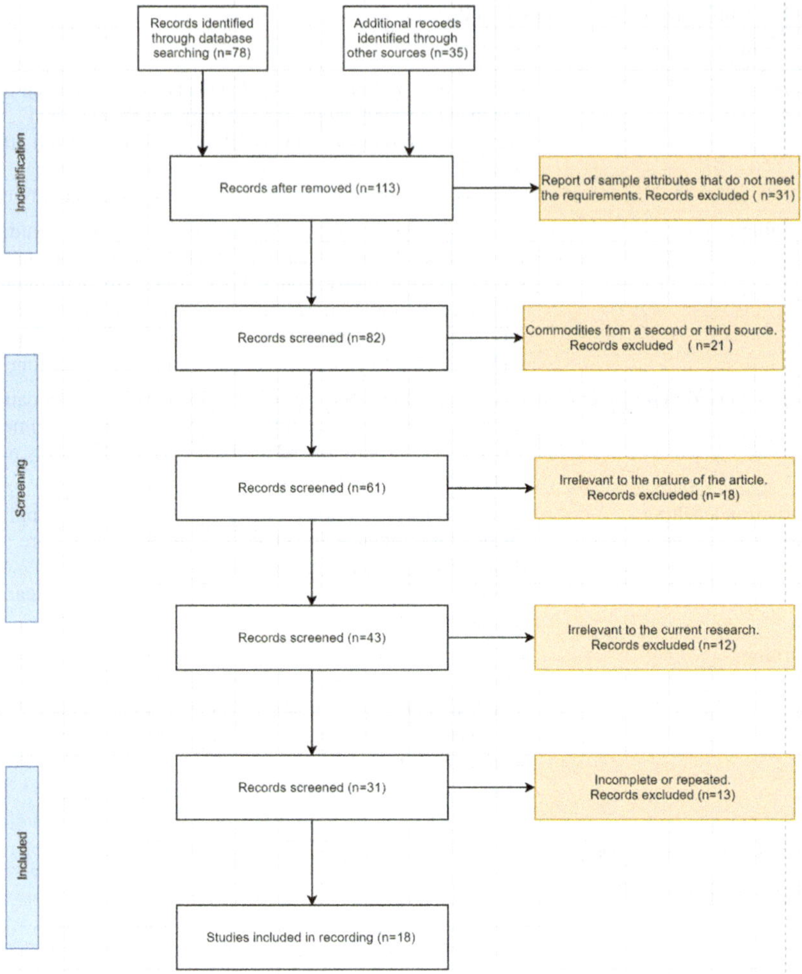

Fig. 3 Inclusion and exclusion criteria as per PRISMA in this book chapter

Table 3 Classified by subject

Classified by subject	
Subject	Number of articles
(1) The impact of artificial intelligence robots on students' creativity and motivation	**7**
(2) The impact of artificial intelligence on education management	11

Table 4 Research data

Author(s) (publication year)	Journal	Title
Alimisis (2009)	ASPETE	Teacher education on robotic-enhanced constructivist pedagogical methods
Barker (2007)	Technology & Education	Robotics as a means to increase achievement scores in an informal learning environment
Bers (2000)	Technology	Technological prayers: Parents and children exploring robotics and values
Erdogan (2013)	Educational Sciences	Defining innovation literacy: Do robotics programmes help students develop innovation literacy skills?
Giannakos (2013)	Educational Sciences	Teaching computer science to young children through creativity: Lessons learned from the case of Norway
Hamner (2013)	Education	Techniques for distributing a STEAM robotics programme through K-12 classrooms
Kahraman (2010)	Technology	Development of adaptive and intelligent Web-based educational systems
Snyder (2019)	Methodology	Literature review as a research methodology: An overview and guidelines
Khanlari (2015)	ICCSE	Using robotics for STEM education in primary/elementary schools: Teachers' perceptions
Kim (2015)	Education	Robotics to promote elementary education pre-service teachers' STEM engagement, learning, and teaching, Computers & Education
Knop (2017)	FIE	A human-interactive robotic programme for middle school STEM education
Lau (1999)	FIE	Creative learning in school with LEGO(R) programmable robotics products
Timms (2016)	Education	Letting artificial intelligence in education out of the box: Educational robots and smart classrooms
Melchior (2005)	Education	More than robots: An evaluation of the first robotics competition participant and institutional impacts
McGrath (2008)	Technology& Education	Build IT: Building middle and high school students' understanding of engineering, science and IT through underwater robotics
Master (2017)	Experimental Child Psychology	Programming experience promotes higher STEM motivation among first-grade girls
UNESCO (2019)	Education	How can artificial intelligence enhance education?

(continued)

Table 4 (continued)

Author(s) (publication year)	Journal	Title
Rus (2013)	Technology & Education	"Recent advances in convert" sectional intelligent tutoring systems

Artificial intelligence educational robots can promote students' learning and teachers' management in teaching. Artificial intelligence robots can recommend a customised learning plan for each scholar based on each student's different preferences and learning methods through big data and cloud computing. They can also help teachers manage the class, for example, using artificial intelligence to answer some learning questions in or after class or collect students' homework completion, substantially enhance teaching efficiency and effectiveness.

6 Conclusion

Artificial intelligence educational robots have a huge impact and are necessary for the education field, as they can help scholars understand abstract concepts in a variety of ways, and at the same time, provide them with timely learning feedback as well as multiple solutions. Scholars who use artificial intelligence educational robots may have a higher level of knowledge and a more active learning attitude towards science and technology.

Furthermore, as the artificial intelligence educational robots often offers an academic learning platform based on Web and online education, they can be "hired" as teachers' colleagues or students' assistants to perform similar functions as teachers and help scholars solve learning problems. Such platforms and tools will substantially improve teaching efficiency and quality and provide students with a better learning experience. Artificial intelligence robots can also be used to customise the learning functions in the platform according to the needs and abilities of students.

In conclusion, artificial intelligence has significantly impacted education, especially in teaching management and autonomous student learning, and should be further applied to develop the future education field.

Acknowledgments This work described in this paper was fully supported by a grant from the Research Grants Council of the Hong Kong Special Administrative Region, China (Project No. UGC/IIDS15/E01/19).

References

Alimisis, D., & Kynigos, C. (2009). Constructionism and robotics in education. *Teacher education on robotic-enhanced constructivist pedagogical methods* (pp. 11–26). School of Pedagogical and Technological Education (ASPETE).

Barker, B. S., & Ansorge, J. (2007). Robotics as means to increase achievement scores in an informal learning environment. *Journal of Research on Technology in Education, 39*(3), 229–243. https://doi.org/10.1080/15391523.2007.10782481.

Bers, M. U. & Urrea, C. (2000). Technological prayers: Parents and children exploring robotics and values. Robots for kids: Exploring new technologies for learning (pp. 193–217). San Diego, CA: Academic Press.

Crowe, D., LaPierre, M., & Kebritchi, M. (2017). Knowledge based artificial augmentation intelligence technology: Next step in academic instructional tools for distance learning. *TechTrends, 61*(5), 494–506.

Eguchi, A. (2014). Educational robotics theories and practice: Tips for how to do it right. *Robotics: Concepts, methodologies, tools, and applications: concepts, methodologies, tools, and applications* (pp. 193–223). IGI Global. https://doi.org/10.4018/978-1-4666-4607-0.ch011.

Erdogan, N., Sencer Corlu, M., & Capraro, R. M. Defining innovation literacy: Do robotics programmes help students develop innovation literacy skills? *International Online Journal of Educational Sciences, 5*(1). 1–9. http://www.yoksis.bilkent.edu.tr/pdf/?doi=12113.

Fang, Y., Chen, P., Cai, G., Lau, F. C. M., Liew, S. C., & Han, G. (2019). Outage limit-approaching channel coding for future wireless communications: Root-proto graph low-density parity-check codes. *IEEE Vehicular Technology Magazine, 14*(2), 85–93.

Giannakos, M. N., Jaccheri, L., & Proto, R. (2013). Teaching computer science to young children through creativity: Lessons learned from the case of Norway. In *Proceedings of the 3rd Computer Science Education Research Conference on Computer Science Education Research* (pp. 103–111). Arnhem, Netherlands.

Hamner, E., & Cross, J. Arts & bots. (2013). Techniques for distributing a STEAM robotics programme through, K-12 classrooms. In *Integrated STEM Education Conference (ISE 2013 IEEE)* (pp. 1–5). Princeton, NJ: IEEE. https://doi.org/10.1109/ISECon.2013.6525207.

Han, J., & Kim, D. (2009). r-Learning services for elementary school students with a teaching assistant robot. In *Proceedings of the 4th ACM/IEEE International Conference on Human Robot Interaction* (pp. 255–256). ACM. La Jolla, CA: ACM/IEEE.

Kahraman, H. T., Sagiroglu, S., & Colak, I. (2010). Development of adaptive and intelligent Web-based educational systems. In *Proceedings of 4th International Conference on Application of Information and Communication Technologies* (pp. 1–5).

Khanlari, A., & Kiaie, F. M. (2015). Using robotics for STEM education in primary/elementary schools: Teachers' perceptions. In *Proceedings of 2015 10th International Conference on Computer Science & Education (ICCSE)*. Cambridge, UK. https://doi.org/10.1109/ICCSE.2015.7250208.

Kim, C., Kim, D., Yuan, J., Hill, R. B., Doshi, P., & Thai, C. N. (2015). Robotics to promote elementary education pre-service teachers' STEM engagement, learning, and teaching. *Computers & Education, 91*(C).14–31. https://doi.org/10.1016/j.compedu.2015.08.005.

Knop, L., Ziaeefard, S., Ribeiro, G. A., Page, B. R., Ficanha, E., Miller, M. H., & Mahmoudian, N. (2017). A human-interactive robotic programme for middle school STEM education. In *Proceedings of Frontiers in Education Conference (FIE)* (pp. 1–7). Indianapolis, IN: IEEE. https://doi.org/10.1109/FIE.2017.8190575.

Kory, J., & Breazeal, C. (2014). Storytelling with robots: Learning companions for preschool children's language development. In *2014 RO-MAN: The 23rd IEEE International Symposium on Robot and Human Interactive Communication* (pp. 643–648). IEEE.

Kory, J. M., Jeong, S., & Breazeal, C. L. (2013). Robotic learning companions for early language development. In *Proceedings of the 15th ACM on International Conference on Multimodal Interaction* (pp.71–72). New York, NY: ACM. https://doi.org/10.1145/2522848.531750.

Krishnamoorthy, S. P., & Kapila, V. (2016). Using a visual programming environment and custom robots to learn c programming and K-12 stem concepts. In *Proceedings of the 6th Annual Conference on Creativity and Fabrication in Education* (pp. 41–48). Stanford, CA: ACM. https://doi.org/10.1145/3003397.3003403.

Lau, K. W., Tan, H. K., Erwin, B. T., & Petrovic, P. (1999). Creative learning in school with LEGO(R) programmable robotics products. In *29th Annual Frontiers in Education Conference, FIE'99* (vol. 2, p. 12D4–26). IEEE. https://doi.org/10.1109/FIE.1999.841676.

Li, R. Y. M. (2018). *An economic analysis on automated construction safety internet of things, artificial intelligence and 3D printing*. Springer.

Li, R. Y. M., & Ng, D. P. L. (2018). Wearable robotics, industrial robots and construction worker's safety and health. In *Advances in Intelligent Systems and Computing* (pp. 31–36).

Master, A., Cheryan, S., Moscatelli, A., & Meltzoff, A. N. (2017). Programming experience promotes higher STEM motivation among first-grade girls. *Journal of Experimental Child Psychology, 160*, 92–106.https://doi.org/10.1016/j.jecp.2017.03.013.

McGrath, E., Lowes, S., Lin, P., Sayres, J., Hotaling, L., & Stolkin, R. (2008). Build IT: Building middle and high school students' understanding of engineering, science and IT through underwater robotics. In *Proceedings of 2008 Annual Conference & Exposition, Pittsburgh, Pennsylvania*. https://www.peer.asee.org/3495.

McKay, M. M., Lowes, S., Tirhali, D., & Camins, A. H. (2015). Student learning of STEM concepts using a challenge-based robotics curriculum. In *Proceedings of 2015 ASEE Annual Conference & Exposition, Seattle, Washington*. https://doi.org/10.18260/p.24756.

Melchior, A., Cohen, F., Cutter, T., Leavitt, T., & Manchester, N. H. (2005). *More than robots: An evaluation of the first robotics competition participant and institutional impacts*. Brandeis University.

Papert, S. (1980). *Mindstorms: Children, computers, and powerful ideas*. Basic Books Inc.

Papert, S. (1993). *The children's machine: Rethinking school in the age of the computer*. Basic Books Inc.

Peredo, R., Canales, A., Menchaca, A., & Peredo, I. (2011). Intelligent Web-based education system for adaptive learning. *Expert Systems with Applications, 38*(12), 14690–14702.

Pokrivcakova, S. (2019). Preparing teachers for the application of AI-powered technologies in foreign language education. *Journal of Language and Cultural Education, 7*(3), 135–153.

Regona, M., Yigitcanlar, T., Xia, B., & Li, R. Y. M. (2022). Opportunities and adoption challenges of AI in the construction industry: A PRISMA Review. *Journal of Open Innovation: Technology, Market, and Complexity, 8*(1), 45

Roll, I., & Wylie, R. (2016). Evolution and revolution in artificial intelligence in education. *International Journal of Artificial Intelligence and Education, 26*(2), 582–599.

Rus, V., D'Mello, S., Hu, X., & Graesser, A. (2013). Recent advances in conver" sectional intelligent tutoring systems. *AI Magazine, 34*(3), 42–54.

Shankar, R., Ploger, D., Nemeth, A., & Hecht, S. A. (2013). Robotics: Enhancing pre-college mathematics learning with real-world examples. In *Proceedings of 2013 ASEE Annual Conference & Exposition* (pp. 117), Atlanta, Georgia. https://www.peer.asee.org/22435.

Sharma, R. C., Kawachi, P., & Bozkurt, A. (2019). The landscape of artificial intelligence in open, online and distance education: Promises and concerns. *Asian Journal of Distance Education, 14*(2), 1–2.

Snyder, H. (2019). Literature review as a research methodology: An overview and guidelines. *Journal of Business Research, 104*, 333–339.

Sutton, H. (2019). Minimise online cheating through proctoring, consequences. *Recruiting Retaining Adult Learners, 21*(5), 1–5.

Timms, M. J. (2016a). Letting artificial intelligence in education out of the box: Educational robots and smart classrooms. *International Journal of Artificial Intelligence and Education, 26*(2), 701–712.

Timms, M. J. (2016b). Letting artificial intelligence in education out of the box: Educational cobots and smart classrooms. *International Journal of Artificial Intelligence and Education, 26*(2), 701–712.

United Nations Education Scientific and Cultural Organization (UNESCO). (2019). How can artificial intelligence enhance education? https://www.en.unesco.org/news/how-can-artificialintelligence-enhance-education.

Wartman, S. A., & Combs, C. D. (2018). Medical education must move from the information age to the age of artificial intelligence. *Academic Medicine, 93*(8), 1107–1109.

Williams, K., Kapila, V., & Iskander, M. G. (2011). Enriching K-12 science education using LEGOs. In *2011 ASEE Annual Conference & Exposition* (pp. 22–630).

Chapter 5
Classification of Construction Accident Court Cases Via Natural Language Processing in Hong Kong

Rita Yi Man Li, Herru Ching Yu Li, Beiqi Tang, and Wai Cheung Au

Abstract Construction accidents often lead to injuries, pain, loss of future earnings and even deaths. One way to lower the likelihood of accidents is to levy high compensation on wrongdoers, including main contractors, subcontractors and even the workers who fail to take safety measures or are careless. Under the common law system, precedents are part of the legal system. Hong Kong is no exception. Therefore, construction companies and legal firms are interested to know the possible amount of compensation. FastText-based classification, a kind of Computer-based automated text classification, classifies documents into predefined categories according to the content of the papers, is proposed in this book chapter for accident compensation in courts. We utilised 3000 sentences in court cases in Hong Kong. 90% of the data was used for training, and 10% was used for testing. The results show that the system's precision for classifying construction accident cases into successfully or unsuccessfully obtained compensation was 95.7%. This demonstrates that the fastText-based classification can successfully classify papers with a high level of accuracy. This pilot research provides a practical example to showcase the possibility of utilising artificial intelligence for predicting the likelihood of obtaining construction accident compensation. This approach could offer a rough estimation of the chance of getting compensation, save human resources, and allow non-specialists without much legal knowledge to have a quick reference on the likelihood of obtaining compensation for accidents. The results can also be generalised to other types of accidents and regions operated under the common law system.

Keywords Accident compensation · Construction accident · Classification · Natural language processing · fastText · Artificial intelligence

---

R. Y. M. Li (✉) · H. C. Y. Li · B. Tang · W. C. Au
Sustainable Real Estate Research Center, Hong Kong Shue Yan University, Hong Kong, China
e-mail: ymli@hksyu.edu

R. T. M. Li et al. (eds.), *Current State of Art in Artificial Intelligence and Ubiquitous Cities*, https://doi.org/10.1007/978-981-19-0737-1_5

1 Introduction

The construction industry plays an essential role in economic development in many countries. Nevertheless, it records five times more workplace deaths than the manufacturing industry (Kanchana et al., 2015; Slates, 2008). Compared to other industries, the construction industry has the highest number of non-fatal injuries or illnesses that requires days off from work (Gambatese & Michael Behm, 2005). It is not only construction workers who are affected by accidents, but also the public, including children (Kanchana et al., 2015; Slates, 2008).

Partly due to construction accidents' severity and frequency, slip and fall incidents on sites have received much attention in research (Kanchana et al., 2015; Nadhim et al., 2016). Li and Poon (2009) proposed that construction workers come across accidents due to a lack of safety knowledge. Szóstak (2019) recognised that workers and participants are exposed to several risk factors relating to injury and death. Li (2019) investigated construction safety informatics, including on-site AI, chatbots, and the IoT, and their possible roles in reducing construction accidents.

The construction industry employs both skilled and unskilled workers exposed to construction site accidents and health risks. Accidents at work in the construction industry results in permanent disability and high death rates (Kanchana et al., 2015; Nadhim et al., 2016). Ensuring workers' health and safety on site is of paramount importance (Li, Ho, & Tang, 2018).

To prevent future fatal building accidents, a comprehensive analysis of the characteristics of fatal accidents is needed to examine their characteristics and make recommendations for accident prevention (Xu & Xu, 2021). Previous research suggested that heat stress profoundly affects physiological reactions, construction injuries, and deaths at work in Oman. In an overview of the safety of construction workers, a model of the causes of accidents shows that safety behaviour in construction and safety awareness are the main factors which affect the likelihood of construction accidents (Hou et al., 2021; Umar & Egbu, 2020). Temporal and spatial conditions influence safety and construction activities in the building environment. Since construction workers spend most of their time on the construction site, they are at high risk of injury and death (Hou et al., 2021; Nadhim et al., 2016).

To enhance construction safety, artificial intelligence is becoming more popular. Li (2018) indicated that the application of AI could be critical to reducing construction accident risks. Li and Leung (2019) researched ways of enabling AI to detect hazards on construction sites. Many experts have pointed out that AI can be helpful for document classification. Thangaraj (2018) argued that AI could play an essential role in the text mining.

In the following sections, Sect. 2 is a literature review of natural language processing, automatic text classification, and compensation and court cases analysis. Section 3 describes the research method and results, and Sect. 4 is the conclusion.

2 Literature Review

2.1 Natural Language Processing

Prior research suggested that the boundaries of natural language processing are not clear (Choudhury & Asan, 2020). Chen (2020) proposed that natural language processing is a generic term that includes techniques that consider language's complexity. Choudhury and Asian (2020) proposed that natural language processing refers to computational capacity, which understands human language by transforming text into machine-readable structured data that can be analysed via machine learning. Schematic representation of natural language processing converts unstructured text into machine-readable structured data that can be analysed by machine learning algorithms. Yse (2019) conceded that natural language processing represents a huge leap forward in machine learning and natural language processing. Sinoara et al. (2017) suggested a wide range of algorithms that could be applied to achieve the goal of natural language processing, such as machine learning.

2.1.1 Uses of Natural Language Processing: Sentiment Analysis

Le Glaz et al. (2021) suggested that natural language processing uses a corporate learning approach that performs well in statistical tasks such as text classification and sentiment mining. Yao et al. (2021) studied Tweets by using SentiStrength. It has a high classification accuracy for informal short texts, which is most suitable for social media content. The built-in linguistic rules allow the computer to label text with neutral, negative and positive valence. The SentiStrength algorithm is based on a lexicon of the words of sentiment. SentiStrength splits it into words, punctuation and emotions when we enter a sentence. Each word will be examined through the lexicon identifying sentiment terms. If the valence, that is, the overall score for a sentence is positive, the tweet is defined as positive emotion. Higher valence indicates a more positive tone. If the valence is negative, the tweet is a negative emotion. A more negative tone is associated with lower valence. Neutral texts have 0 value (Table 1).

2.1.2 Automatic Text Classification

Classification is the labelling procedure according to predefined criteria, for example, classifying an album according to its music genre. In the data mining context, classification consists of data analysis in two steps: (1) induces a model from a set of training tuples by using machine learning and statistical algorithms that can (2) predict which class of a document that we have not trained before for testing the accuracy (Stein et al., 2019).

Automatic text classification is useful because it reads the document's content and classifies documents according to specific predefined parameters determined by the

Table 1 Examples of negative, neutral and positive quotations of cif_ireland tweets (Yao et al., 2021)

Valence	Examples
−4	Male site workers in are three times more likely to complete suicide than UK male. The easiest thing that we can all do is talk. #CIFSafety20 https://t.co/Z5uQBPjS28
−3	Frightening statistics… of the 352 suicides in Ireland in 2018.. 80% were male… mind your #CIFSafety19 https://t.co/YHNjFwbNj2
−2	We feel it's our to raise awareness; encourage open; honest when it comes to suicide. Male workers are 6 times more likely to die from suicide than from falling from height. As a global we must unite to help #SuicidePrevention https://t.co/frYm20QIaD
−1	Construction sites can be very dangerous places, but the right can do a lot to mitigate that risk. PPE is not the end there are quite a few that are necessary to protect workers, key is your risk process. https://t.co/CjvCNFu0bo https://t.co/1mAkNOZsmX
0	Our 1 day course in Temporary Works General Awareness is designed to enhance site by equipping workers with the necessary to ensure safe operations on site at all times, resulting in fewer risks accidents in. https://t.co/1VMHXAVRyN https://t.co/tvFZX1 dgb5
1	Think! Just goes to show how important it is to wear your gear. #SafetyFirst #ThinkSafety https://t.co/qCR71znimq
2	We had a great first of Construction Industry Federation's Safety Week. Here's some pictures from our sites, including a water rescue at Dublin Port Company. #CIFSafety18 https://t.co/JtH8x4TDe6
3	Very excited to be at the @CIF_Ireland Health; Safety conference in #crokepark, #Dublin with @ems4safety. We will be running some demonstrations of our safety, if you're attending come and say! #CIFSafety19 #healthandsafety #safety #software https://t.co/F5KYEZrry4
4	Agreement between speakers, attendees and @ddccarey as he speaks to the panel that it is hugely positive that the of is now prominent in construction health; safety conversations. #cifsafety https://t.co/vq0ncd3dD4

paper, saving hours of manual work. In the case of a systematic check, the algorithms are optimised to find relevant data sets and the most accurate model. What is unique about systematic verification is that not all relevant data sets (e.g. titles, abstracts) need to be viewed by the researcher and a variety of concepts need to be learned, which requires flexibility in the modelling approach and careful error evaluation (Oliver, 2021; van de Schoot et al., 2021).

Machine learning algorithms can be trained to classify relevant documents, saving reviewers dozens of hours. Many cluster algorithms have been proposed to group data in machine learning and data science literature. This technique, also known as non-parametric density clustering, separates high-density clusters from low-density clusters and can be used for model building (Oliver, 2021; Sarker, 2021).

In the following section, we will discuss various applications based on machine learning. NLP algorithms are classified as a kind of machine learning because they learn while processing relevant data and are able to perform their tasks better when they process additional training data. The information extraction used in NLP includes name, entity, and relation extraction (also known as association extraction) (Oliver, 2021; Sarker, 2021).

Kamruzzaman (2020) stated that text classification was the process of classifying documents into predefined categories based on their content. The process should assign the natural language text to the various categories automatically. Text classification is a primary requirement of text retrieval systems—which retrieve texts in response to a user query. Such a system may transform the text in some way, such as producing summaries, answering questions or extracting data. Linss (2019) described an AI-based tool that classified results according to their relevance, determined how often an essay (n this case) had been cited, and displayed a visual representation of the relationships between studies. Mahoney, Zhang et al. (2019) determined that applying AI text classification could make it much more efficient, and it could cull massive volumes of data and identify responsive documents for use in these matters.

2.1.3 Natural Language Processing for Retrieving and Extracting Data and Text Analytics

Modern electronic text processing systems are ubiquitous—from instant messaging apps in smart phones to virtual repositories. All these impose considerable challenges to address users' new information needs. One endeavour automatically classifies textual data so that information system users can more easily retrieve, extract, and manipulate information to recognise patterns and generate knowledge (Stein et al., 2019).

Barba (2019) noted that natural language processing is used for text analytics to identify the most relevant words and phrases in a text, such as "hello," "good", and "bad." Such approaches are examples of supervised machine learning. NLP may also be implemented as an algorithm that extracts meaning from large data sets, and such a method is known as unattended machine learning.

2.1.4 Software for Natural Language Processing

There has been more research on natural language processing and applied it in various areas in recent years. For example, Le Glaz et al. (2021) noted that natural language processing offers new perspectives in mental health research to support clinical practice. However, ethical issues must be considered before taking it as an additional tool for mental health care. Thalia, a machine learning system based on CRFs, indexes new PubMed articles by genes, chemicals, diseases, drugs, metabolites, proteins, species, and anatomical units, allows an automatic update of the index and provides a user interface to interact with the identified concepts. In October 2018, PubMed adopted a hybrid approach, using the PubMed Medical Text Indexer (MTI) system to assign articles to a mesh of terms and then use a combination of machine learning to create rules for giving terms to human interventions (Marshall & Wallace, 2019).

Steven Bird (2019) indicated that there was often a need to solve several classification problems that were closely linked simultaneously. To obtain the dependencies between related classification tasks, we can apply standard classifiers and models that choose a suitable label for each collection and related input. Sequence classification is another methodology, and one sequence classification strategy is sequential classification or greedy sequence classification. This proceeds to find the best labels for the text input and then apply the answer to find a suitable label for the following input; labels can be anything, "good" and "bad" are often used. Regarding natural language processing used for document classification, Biswas (2018) argued that many tools could help, such as MALLET and Apache Open NLP.

2.2 Construction Accident Compensation Court Cases Analysis

Murphy (2020) advised that it is vital to work with an experienced lawyer who understands the various factors that affect injuries' compensation. Nevertheless, setting a reasonable compensation level for injured workers can be complex. Multiple stakeholders, including the employer, contractors, subcontractors and other parties involved in the construction project, may be liable for an accident. Barsumian (2020) believed that injury attorneys could file lawsuits against negligent makers, owners and maintainers of equipment, property owners who did not disclose their knowledge of hazards, and other construction companies who did not adhere to the best safety practices construction sites. Hiring an experienced personal injury lawyer who knows how to identify the causes of an injury, such as negligence, is the first step in getting the accident compensation.

Gürcanli and Müngen (2013) indicated that if a person's negligence leads to an avoidable building accident, that person should be held liable for the injury to the victim. It is always important to look at who is to blame for building accidents. A safety expert is responsible for determining whether there is a link between the safety standards of a construction site and the risk of injury of accident victims. Sometimes, the architect and/or engineer may be liable. Miller (2020) determined that in some construction accident court cases, the system architect and/or the engineer responsible for designing systems supporting the design, construction, maintenance and operation of buildings could be held liable for their negligence or improper use of materials/equipment.

3 Research Method and Results: Classifying Legal Document Using Artificial Intelligence

In this research, we have utilised fastText, a system developed by Facebook, to classify legal documents relating to compensation payments. The fastText system has a trained English language vector model derived from Wikipedia, Common Crawl and other sources. Every word is stored along with several different tokens, and the morphology and word representations are replaced with vectors. Unlike word2vec and other word vector models, FastText has implemented n-grams of characters (Agibetov et al., 2018). Words can be used easily to discover unknown vector representations because the length, n, is variable. The model can also be adapted to both supervised and unsupervised learning.

FastText has been utilized in previous research. Examples include Stein et al. (2019) stated that efficient distributed numerical word representation word embedding models combined with machine learning algorithms had enhanced automatic document classification tasks in recent times. Nonetheless, the effectiveness of these techniques has not been assessed for the hierarchical classification of texts. They trained classification models with machine learning algorithms, Keras' CNN, XGBoost, fastText, SVM, and word embedding generation methods such as GloVe, word2vec, and fastText publicly available data and evaluated them with measures for the hierarchical context. FastText reached a lcaF1 of 0.893 on a labelled version of RCV1 data. Their search results confirm that wordboarding is a promising approach for HTC.

We implemented supervised learning: each sentence was represented as a bag of grams and was normalised to keep the local word order the same (Kuyumcu et al., 2019). In this paper, the author uses 3000 sentences as data. The training split is 90% for training and 10% for testing. The results show that the system's precision for classifying construction accident cases is 95.7% and that the recall is 95.7%. This demonstrates that a fastText classification can successfully classify papers as relevant to accepting or rejecting a compensation case at a reasonably high accuracy rate.

Sentence in document	Confidence Percentage (%)	Document Name
"G. QUANTUM OF DAMAGES The plaintiff has received employees' compensation of $833,322.51."	98.06	*Cheung Chung v. Softtrans Supply Chain [2018]*
"PSLA was awarded $50,000 for the first accident and $150,000 for the second one."	85.99	*Cheung Chung v. Softtrans Supply Chain [2018]*
"PSLA was agreed at $150,000 between the parties."	90.90	*Cheung Chung v. Softtrans Supply Chain [2018]*
"The Court awarded $84,000 as PSLA."	99.97	*Cheung Chung v. Softtrans Supply Chain [2018]*

(continued)

(continued)

Sentence in document	Confidence Percentage (%)	Document Name
"I would award $100,000, which is in my view reasonable, under this head."	99.42	*Cheung Chung v. Softtrans Supply Chain [2018]*
"The defendant accepts $5,000 to be awarded under this head."	99.93	*Cheung Chung v. Softtrans Supply Chain [2018]*
"What I would do here is simply to retain the sum of $3,000 to cater to all T's extracurricular activities, and this sum would include Olympiad Math as well."	87.34	ESBU v LCH [2013]
"However, as T will be travelling during peak season, I think the sum of $19,000 including the airfares and all other expenses is about right."	95.75	ESBU v LCH [2013]
"As I said in paragraph 39 above, I would allow a monthly sum of $3,000 for all of T extracurricular activities, including the Olympiad Math which the mother will pay in the future."	97.31	ESBU v LCH [2013]
"I accept this being the actual monthly expenses of T in this item. Medical Expenses at $2,100[31] The wife now claims that T monthly medical expenses amount to $2,100, which is made up by the sum of $200 as general medical expenses and $1,900 being consultation fee with a psychologist called Annie Ho. [32] For general medical and dental expenses, I understand that both the mother and the father employers would provide medical coverage for their employees and children."	74.40	ESBU v LCH [2013]
"As to quantum, I propose to assess the further costs summarily at $2,500."	98.13	ESBU v LCH [2013]

(continued)

(continued)

Sentence in document	Confidence Percentage (%)	Document Name
"Out of this sum, it would be reasonable to attribute a sum of $2,000 being the father personal travelling expenses, and the balance of $3,000 being travelling expenses under the general expenses for the whole family."	51.64	ESBU v LCH [2013]
"For the period April 2000 to March 2003, I assess the expenses as $841,475.00 per annum and net income thus $486,289.00 per annum before tax, for which I shall allow a deduction of 16%."	70.81	Young vs Lee Chu [2003]
"The plaintiff has claimed, and the defendant conceded, an amount of $281,750.44 under this head, which I award."	99.52	Young vs Lee Chu [2003]
"The defendant has offered an amount of $10,000.00: if he is prepared to pay that sum, I am prepared to order."	84.98	Young vs Lee Chu [2003]
''Once again the defendant has offered the sum of $10,000.00, which appears reasonable to me."	98.52	Young vs Lee Chu [2003]

4 Conclusion

This research explores the possibility of using fasText AI, a system developed by Facebook, to classify construction accident court cases involving compensation. The pilot test demonstrates that AI classification can provide highly accurate results. This would benefit any law firm whose work involves scanning many documents to identify relevant court cases, construction companies when they need to search for compensation cases with specific attributes, and academic researchers undertaking compensation research.

According to the National Institute of Standards and Technology (NIST), while standardised data make training data and machine learning inputs available and usable for authorised users, artificial intelligence trained on unstructured or biased data can produce misleading results (Choudhury & Asan, 2020). As such, further research is needed and might provide beneficial effects.

Acknowledgements The work described in this paper was supported by a grant from the Research Grants Council of the Hong Kong Special Administrative Region, China, Project no. UGC/FDS15/E01/18

Funding Information This book chapter is an extension of a prior shorter paper which was presented and published in ACM Proceedings:

Rita Yi Man Li, Herru Ching Yu Li, Beiqi Tang and Wai Cheung Au 2020. Fast AI classification for analysing construction accidents claims. In Proceedings of ACM. ACM, New York, NY, USA, 4 pages. https://doi.org/10.1145/1234567890.

References

Agibetov, A., Blagec, K., Xu, H., & Samwald, M. (2018). Fast and scalable neural embedding models for biomedical sentence classification. *BMC Bioinformatics, 19*(1), 541. https://doi.org/10.1186/s12859-018-2496-4

Barba, P. (2019). Machine learning for natural language processing. Retrieved from https://www.lexalytics.com/lexablog/machine-learning-natural-language-processing.

Barsumian. (2020). Personal injury lawyers. Retrieved from https://www.barsumianlaw.com.

Biswas, S. (2018). Let's talk about natural language processing. Retrieved from https://dzone.com/articles/lets-talk-about-natural-language-processing.

Chen, P.-H. (2020). Essential elements of natural language processing: What the radiologist should know. *Academic Radiology, 27*(1), 6–12. https://doi.org/10.1016/j.acra.2019.08.010

Choudhury, A., & Asan, O. (2020). Role of artificial intelligence in patient safety outcomes: Systematic literature review. *JMIR Medical Informatics, 8*(7), e18599. https://doi.org/10.2196/18599

Gürcanli, G. E., & Müngen, U. (2013). Analysis of construction accidents in turkey and responsible parties. *Industrial Health, 51*(6), 581–595. https://doi.org/10.2486/indhealth.2012-0139

Hou, L., Wu, S., Zhang, G., Tan, Y., & Wang, X. (2021). Literature review of digital twins applications in construction workforce safety. *Applied Sciences, 11*(1), 339.

Gambatese, J. A., & Michael Behm, J. H. (2005). *Investigation of the viability of designing for safety*. Retrieved from http://elcosh.org/document/1892/d000693/investigation-of-the-viability-of-designing-for-safety.html.

Kamruzzaman, S. M. (2020). Text classification using artificial intelligence. *Journal of Electrical Engineering, 33*, 1–6.

Kanchana, S., Sivaprakash, P., & Joseph, S. (2015). Studies on labour safety in construction sites. *The Scientific World Journal, 2015*, 590810. https://doi.org/10.1155/2015/590810

Kuyumcu, B., Aksakalli, C., & Delil, S. (2019). *An automated new approach in fast text classification (fasttext): A case study for turkish text classification without pre-processing*. Paper presented at the 3rd International Conference on Natural Language Processing and Information Retrieval, Tokushima.

Le Glaz, A., Haralambous, Y., Kim-Dufor, D.-H., Lenca, P., Billot, R., Ryan, T. C., … Lemey, C. (2021). Machine learning and natural language processing in mental health: Systematic review. *Journal of Medical Internet Research, 23*(5), e15708. https://doi.org/10.2196/15708.

Li, R. Y. M. (2018). *An economic analysis on automated construction safety*. Springer.

Li, R. Y. M. (2019). *Construction safety informatics*. Springer.

Li, R. Y. M., Ho, D. C. W., & Tang, B. (2018). *Factors which affect construction safety in different types of construction work*. Paper presented at the Advances in Safety Management and Human Factors, Cham.

Li, R. Y. M., & Leung, T. (2019). Computer vision and hybrid reality for construction safety risks: A pilot study. *Advances in Intelligent Systems and Computing, 1027*.

Li, R. Y. M., & Poon, S. W. (2009). Future motivation in construction safety knowledge sharing by means of information technology in hong kong. *Journal of Applied Economic Sciences, 4*(9).

Linss, M.-A. (2019). Looking ahead: The potential for ai in literature search. Retrieved from https://info.reprintsdesk.com/about/blog/2019/looking-ahead-the-potential-for-ai-in-literature-search.

Thangaraj, M. S. (2018). Text classification techniques: A literature review. *Interdisciplinary Journal of Information, Knowledge, and Management, 13*, 117–135. https://doi.org/10.28945/4066

Mahoney, C. J., Zhang, J., Huber-Fliflet, N., Gronvall, P., & Zhao, H. (2019). *A framework for explainable text classification in legal document review*. Paper presented at the 2019 IEEE International Conference on Big Data (Big Data).

Marshall, I. J., & Wallace, B. C. (2019). Toward systematic review automation: A practical guide to using machine learning tools in research synthesis. *Systematic Reviews, 8*(1), 163. https://doi.org/10.1186/s13643-019-1074-9

Miller. (2020). Accident lawyer. Retrieved from https://atlantaadvocate.com/georgia/atlanta/construction-accident-lawyer/.

Murphy, B. O. (2020). Nyc construction accident lawyers on your side. Retrieved from https://www.blockotoole.com/Construction-Accident-Help/.

Nadhim, E. A., Hon, C., Xia, B., Stewart, I., & Fang, D. (2016). Falls from height in the construction industry: A critical review of the scientific literature. *International Journal of Environmental Research and Public Health, 13*(7), 638. https://doi.org/10.3390/ijerph13070638

Oliver, G. (2021). Nlp-aided systematic literature review: Why it's needed and how it works. Retrieved from https://towardsai.net/p/nlp/nlp-aided-systematic-literature-review-why-its-needed-and-how-it-works.

Sarker, I. H. (2021). Machine learning: Algorithms, real-world applications and research directions. *SN Computer Science, 2*(3), 160. https://doi.org/10.1007/s42979-021-00592-x

Sinoara, R. A., Antunes, J., & Rezende, S. O. (2017). Text mining and semantics: A systematic mapping study. *Journal of the Brazilian Computer Society, 23*(1), 9. https://doi.org/10.1186/s13173-017-0058-7

Slates, K. (2008). The effects of leadership in the high hazard construction sector: Injuries and fatalities an issue of leadership and not hazard. *Leadership and Management in Engineering, 8*(2), 72–76. https://doi.org/10.1061/(ASCE)1532-6748(2008)8:2(72)

Stein, R. A., Jaques, P. A., & Valiati, J. F. (2019). An analysis of hierarchical text classification using word embeddings. *Information Sciences, 471*, 216–232. https://doi.org/10.1016/j.ins.2018.09.001

Steven Bird, E. K. a. E. L. (2019). Learning to classify text. In *Natural language processing with python*.

Szóstak, M. (2019). Analysis of occupational accidents in the construction industry with regards to selected time parameters. *9*(1), 312. https://doi.org/10.1515/eng-2019-0027.

Umar, T., & Egbu, C. (2020). Heat stress, a hidden cause of accidents in construction. *Proceedings of the Institution of Civil Engineers - Municipal Engineer, 173*(1), 49–60. https://doi.org/10.1680/jmuen.18.00004.

van de Schoot, R., de Bruin, J., Schram, R., Zahedi, P., de Boer, J., Weijdema, F., … Oberski, D. L. (2021). An open source machine learning framework for efficient and transparent systematic reviews. *Nature Machine Intelligence, 3*(2), 125–133. https://doi.org/10.1038/s42256-020-00287-7.

Xu, Q., & Xu, K. (2021). Analysis of the characteristics of fatal accidents in the construction industry in china based on statistical data. *International Journal of Environmental Research and Public Health, 18*(4), 2162. https://doi.org/10.3390/ijerph18042162

Yao, Q., Li, R. Y. M., Song, L., & Crabbe, M. J. C. (2021). Construction safety knowledge sharing on twitter: A social network analysis. *Safety Science, 143*, 105411. https://doi.org/10.1016/j.ssci.2021.105411

Yse, D. L. (2019). Your guide to natural language processing (nlp). Retrieved from https://www.kdnuggets.com/2019/05/guide-natural-language-processing-nlp.html.

Chapter 6
Smart Home, Robots, and Robotic Arms for Elderly and Disabled Persons: A Lab Experiment Research Agenda

Rita Yi Man Li

Abstract The Chief Executive's Policy addresses direct Hong Kong moving towards a smart city. Although the smart home is one of the major features in smart city, understanding on smart home is limited among lay people in Hong Kong. Elderly and disabled persons are no exception. The smart home gadgets are useful among these two special groups. For example, a water sensor installed in the washing basin can inform the elderly who forget to turn off the water closet, carbon monoxide sensor alerts the resident if there is a gas leakage problem. Bed sensor can send messages to care takers' apps when the elderly is suspected to be fainted in the bed. The smart home robots companion lonely elderly who may not have companions with them. Video recording system can inform the caretakers in case of fall incidents so that appropriate and timely action can be taken. On the other hand, a voice controllable system can help disabled persons control the electrical appliances such as television, and play music in the whole room by using their voice rather than their hand. It helps the blind people and those who cannot move around the room easily or have limited hand function. Robotic arms on wheelchairs help disabled persons with limited hand function to grab things on the floor, eat and drink. Facial recognition systems prevent strangers enter the home via facial recognition systems. This research agenda proposes to (1) build a sample smart home in a University to demonstrate the possible future direction in smart home. It incorporates three generations of the smart home gadgets (a) the smart home gadgets which are connected by a network system; (b) smart gadgets which have a certain level of artificial intelligence which can reply to our questions; (c) smart home robots which can move around the home to move around the home quickily on behalf of the elderly and disabled persons. (2) Install smart home gadgets at an affordable price for 200 households with elderly/disabled persons to inform the elderly and disabled persons about the possibility of these tools helping their daily lives. Practically speaking, the project can provide valuable insights to policy makers when they implement smart city, elderly, and disabled persons' policies. Caretakers can know more about the current-state-of-art in smart home gadgets, which may reduce their everyday work. This can also raise the awareness of the elderly and disabled persons' needs in Hong Kong.

R. Y. M. Li (✉)
Department of Economics and Finance, Sustainable Real Estate Research Center, Hong Kong Shue Yan University, Hong Kong, China

R. T. M. Li et al. (eds.), *Current State of Art in Artificial Intelligence and Ubiquitous Cities*, https://doi.org/10.1007/978-981-19-0737-1_6

Academically speaking, the smart home gadgets installed in the smart home laboratory shall provide important information to university researchers to study people's behaviour/reaction to the smart home. Whilst we can find many of the smart home laboratories in overseas universities such as the University of West Florida and Iowa State University, we cannot find any smart home laboratory designed for elderly and disabled persons. This research agenda will fill the research gap.

Keywords Ageing population · Smart home · Robot · Disabled persons

1 Introduction

The ageing population is a Worldwide phenomenon at present. People over the age of 65 are the fastest-growing age group worldwide. In 2019, they surpassed the number of children under 5 for the first time in human history (Jarzebski et al., 2021). It is anticipated that an increase in population age invariably leads to an escalating demand for various health services among older adults (Sin et al., 2015). On the other hand, if older adults can optimize their health and live in an environment that promotes their active participation, their experience, skills, and wisdom will be a valuable resource for society (Robson et al., 2006). Therefore, to safeguard the health and functional ability of older adults are of top priority for ageing society in any place around the globe, Hong Kong is no exception. The smart home technology enhances seniors' independency and quality of life. Without a doubt, smart homes offer solutions for the elderly with a disability to live more independently (Chan et al., 2009), allowing them to take control—turn on and off lights, figure out who knocks on the door from the mobile phones. It can also showcase the collaboration between Academia and industry.

This book chapter provides a research agenda and conceptual ideas of smart home research for the elderly. It aims to promote the importance of bringing in new technologies to create better quality of life for seniors and persons with disabilities. Whist the Policy address 2021 has already identified the importance of making our city smarter (The Hong Kong Special Administrative Region of the People's Republic of China, 2021) and many of us aware of big data etc., understanding on how smart home can improve the quality of life among these two groups remains limited. As they have different pros and cons, we can also use them in the smart home lab to examine under which situations they will better serve some elderly or disabled persons.

In this research agenda, we propose providing smart home gadgets to 100 elderly and 100 disabled persons at affordable price, i.e., HK$15,000 for each household. Two robots are included mainly based on their ability as mentioned above, i.e., Pepper and Nao in the smart home in a university lab. A robotic arm will be installed in the wheelchair to help the disabled persons grab things on the floor, feed, and drink, open and close the fridge and grab things inside, open and close the doors, use a microwave etc.

2 Literature Review

The Internet-of-Things links objects wirelessly to a network that enables data sharing; network AI simultaneously analyses IoT data for making decisions such as the temperature of the air conditioners. The smart home is one example of the application for this technology (Yigitcanlar et al., 2021). Smart homes are stuffed with appliances that can be controlled remotely. They are heralded as the wave of the future. They provide important solutions for the disabled persons. A man who has trouble bending can use smartphones to turn on and off the electrical appliances. A blind person can voice-activated the TV guide for changing different channels. With a doubt, smart homes offer solutions for disabled people to live independently, allowing them to take control—turn on and off lights, figure out who knocks on the door from a mobile phone. Using an app in mobile phone can turn the lights on and off, dim or change the colors (Gupta, 2015).

The smart home is often linked to the concept of home automation. It delivers commands over a ubiquitous wireless network to electrical devices, gadgets and various applications. It enables users to control temperature, lighting, ventilation and protects occupants against intruders. The Bluetooth app saves energy, improves consumer comfort by reducing energy demand during peak periods and allows more electrical appliances to be operated simultaneously (Li et al., 2016).

The first generation of smart homes enables occupants to operate electrical appliances according to a pre-defined model. For example, advanced metering infrastructure sends electricity price messages to residents' smart meters, and meters adjust their consumption accordingly. ZigBee-based home automation enables the communication between electrical devices and the Internet, so residents can use it to control electrical devices with an app. Therefore, the smart home provides support to patients with disabilities who cannot quikily move around the home (Li et al., 2016).

The second generation of the smart home enhances the traditional home automation system through various forms of artificial intelligence such as voice control that can enhance residents' comfort. Various computation methods are used for automation. In 2014, Amazon Echo can listen to user questions/orders and answer/answer by turning on electrical devices, playing music, etc. It reports traffic and weather, provides user updates on user news and schedules. It processes the ability to learn and becomes more intelligent when users engage with it. In the iSpace of the University of Essex, it is equipped with smart devices to detect and learn user models and provide suggestions based on user needs and requirements (Li et al., 2016).

The third generation of Smart Home utilizes a robot companion that can interact with humans as a family member. For example, they can connect to smart home devices, move around independently at home, take pictures, make video telephone calls (Li et al., 2016).

2.1 Smart Home and Robotics for the Disabled Persons

Most companies that make smart home products do not specifically market them to people with disabilities. Although Nest (owned by Google) dodged answering questions about whether people with disabilities are buying smart home products, it conceded that it had received input from blind people during product design. AT&T and Comcast have plans to create technology specifically for people with trouble in moving around the house (Gupta, 2015).

On the other hand, a few smart robots address the needs of the disabled. For example, the attachment of a robotic arm to a wheelchair increases the user's independence: they may then be able to open the door by themselves with the help of the robotic arm. Other rehabilitation robots assist disabled individuals in their everyday's needs such as eating, drinking, and assisting office work (Stefanov et al., 2004).

2.2 Smart Home Appliances for the Elderly

According to the 2011 Population Census, there were 941,312 older persons in Hong Kong in 2011, constituting 13.3% of the whole population in Hong Kong. The number of older persons increased at an average annual growth rate of 4.8% over the past 50 years (from 1961 to 2011), compared to that of the whole population at 1.6% over the same period. Among older persons in 2011, 12.7% of older persons living alone, 23.6% with spouse only. When we compare that with the distribution ten years ago, the proportion of older persons living with children (including those with spouse and children and those with children only) dropped from 56.8% in 2001 to 51.2% in 2011. On the contrary, the proportions of older persons lived with spouse only and those living alone rose by 5.2 and 1.4 percentage points respectively over the same period.

This situation presents both challenges and opportunities to the existing health care system. It is anticipated that the prevalence of common chronic diseases will increase as population age invariably leads to an escalating demand for various health services among older adults. As such, the smart home serves to support community-based patient care. Using .net in Scotland, for example, enables physicians and family members of seniors to see the status of seniors in real-time (Li, 2013). Other examples are listed in the following table (Table 1).

Table 1 Key issues facing seniors and how intelligent home gadgets can help them (Gaddam et al., 2010; Mutual Mobile, 2015)

Major problems faced by the elderly	How smart home gadgets provide solutions to the elderly
Emergency response	Smart home gadgets provide a panic button that immediately notifies the relatives or 999 in case of emergency like falling at home
Eyesight and hearing	Enhanced alarms in doorbells, smoke detectors and appliances are more effective in alerting a person with a visual or hearing impairment in the event of incidents at home
Memory	When mental acuity fades with age, forgetting may lead to serious safety issues. The smart home reminds us to turn off a coffee maker, lock doors automatically and even take medicine every day
Video monitoring	Caregivers can see seniors by watching video inside and outside of residential units
Medication assistance	The smart system provides advice on medications and dosage when they should be taken, locks out inappropriate medications when medications should not be taken at that time

3 Proposed Smart Home Lab and Smart Home Gadgets for the Elderly and Disabled Persons

The project includes two parts. Part one will set up a smart home lab in a University to demonstrate the latest smart home equipment. However, some of these gadgets may be too expensive in the eyes of public. For example, bed sensors (to detect if the elderly does not have any motion in bed for a long time), robots and robotic arms. Hence, we will only consider them as a futuristic smart homes. In part 2 of the projects, we will provide smart home gadgets to 100 elderly and 100 disabled persons at affordable prices, i.e., HK$15,000 for each household. These can provide an excellent opportunity to let these laymen know more about smart home gadgets (first and second generations of smart home products only). The following figures illustrate the proposed smart home design in university (Figs. 1, 2, 3 and 4).

Note

1. The bed pressure sensor is used to detect the elderly, whether they still have motion in bed after sleeping for some time. If the sensor detects some abnormal situation, that will send messages to the caretakers.
2. Medication reminder is used to remind the elderly to take medicine.

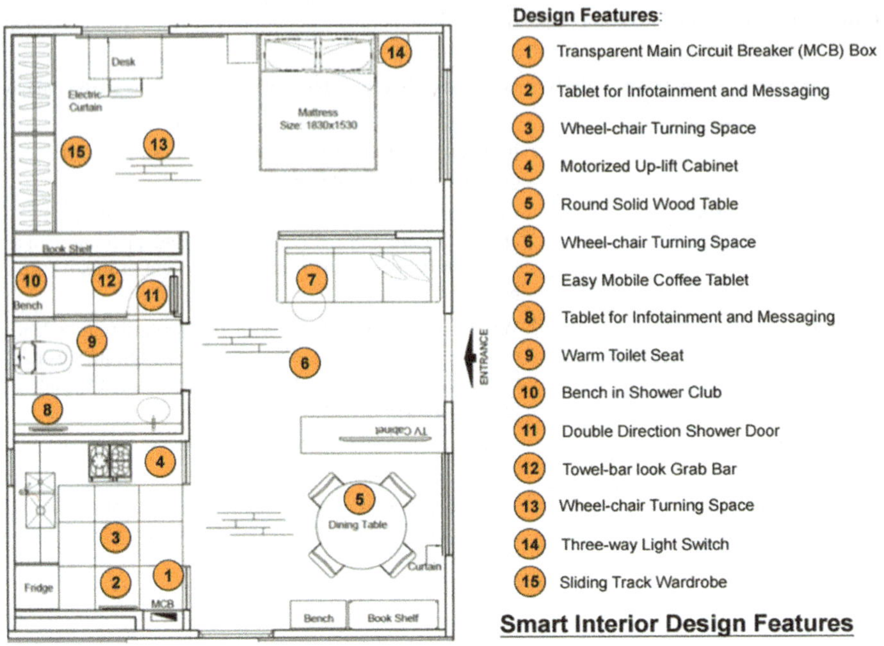

Fig. 1 Smart home features for the disabled in particular (6 and 7). Smart Interior Design Features (detail explanation for the use of the gadgets in this figure can be found in Appendix 1)

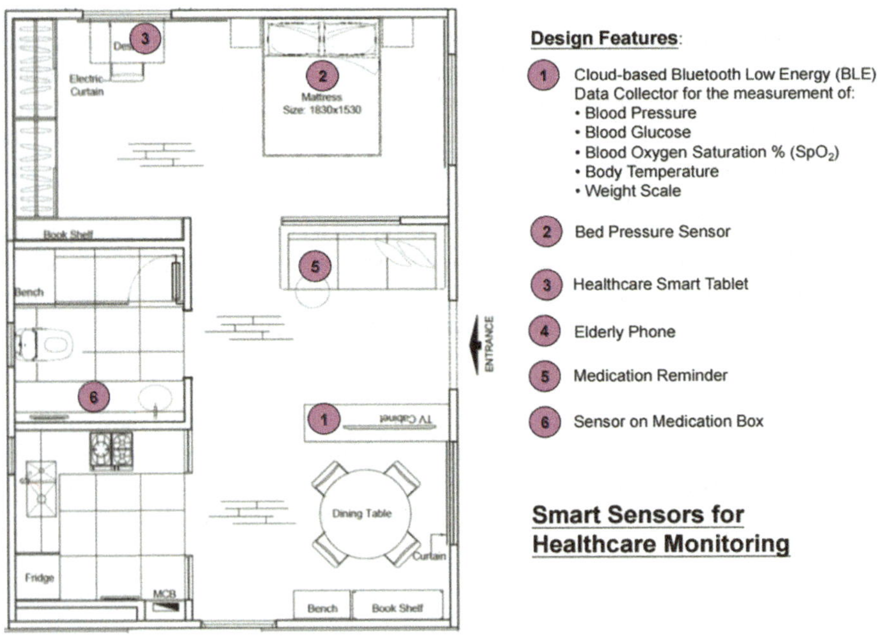

Fig. 2 Smart sensors for monitoring health of the elderly

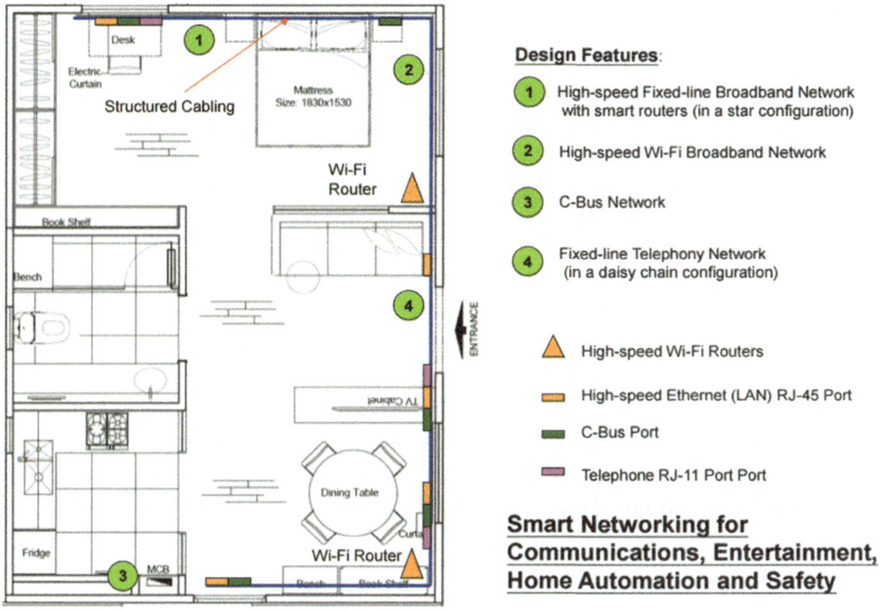

Fig. 3 Smart network for home automation, communications and entertainment

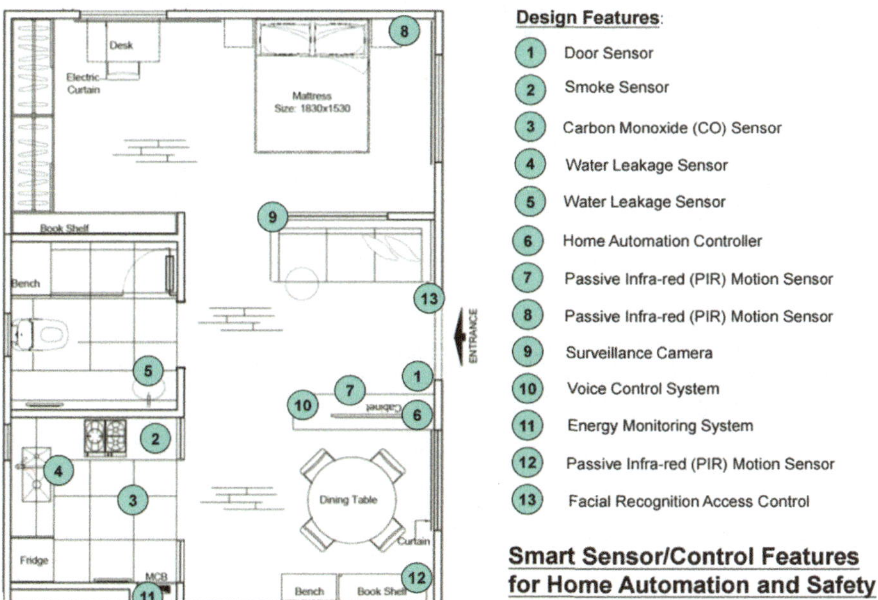

Fig. 4 Smart sensor/control features for home automation and safety: #4&5 water leakage sensor, #9 surveillance camera—allow caretakers to view, #10 voice control system—allow the blind to control the home automation appliances, #13 facial recognition access control

4 Smart Home Robots

Smart home robots in this project include smart home robots and robotic arms. In the modern smart home, smart home robots can act as a companion to the elderly. They can talk to the elderly, help the caretakers to walk close to the elderly and report whether the elderly has fallen. The elderly can also instruct the robots to go to some of the rooms on their behalf. They can also demonstrate to the elderly how to do exercise in the correct position.

In this project, two robots are included mainly based on their ability, i.e., Pepper and Nao in the smart home in Shue Yan. Pepper is larger and can move from one room to another easier than Nao. However, Nao has been equipped with the software that can do exercises with the elderly. Besides, smaller flats may fit better with Nao as the size is smaller. Nevertheless, the relatively small size of Nao also implies that moving from one place to another takes a lot longer time than Pepper. Both can be used to control smart home gadgets such as lighting systems etc. The detail of the design of these two robots is shown in the following figures (Figs. 5, 6, 7 and 8).

Pepper - Technical overview

Dimensions

Arm down

Height (mm)	Width (mm)	Depth (mm)
1208.5	477.2	424

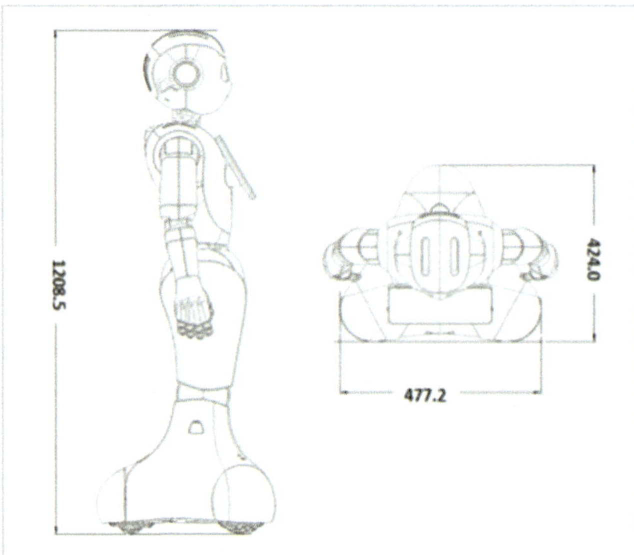

Fig. 5 Size of pepper

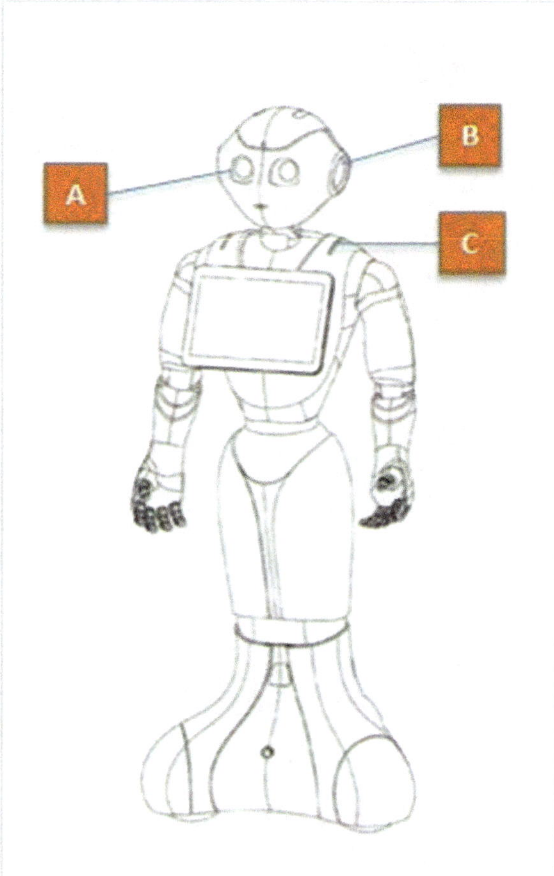

Part	Name	Qty	Description
A	Eyes	2×8	RGB FullColor
B	Ears	2×10	16 Blue levels
C	Shoulders	2x2	RGB FullColor

Fig. 6 Parts of pepper robot

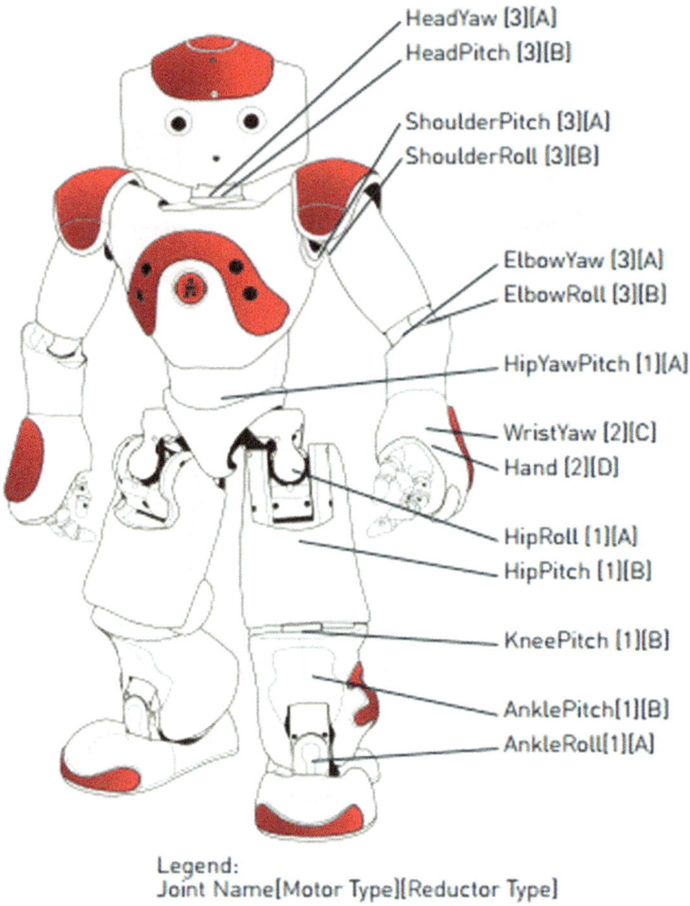

Fig. 7 Design of Nao robot (Softbank 2017). *Note* Nao can be put on the table and hence the size is smaller than Pepper

5 Robotic Arm

The robotic arm will be installed on the wheelchair to help disabled people to grab things on the floor, feed and drinks, open and close the fridge and grab things inside, open and close the doors, use a microwave etc (Fig. 9).

Functions	Details
Moving	25 degrees of freedom and a humanoid shape enable him to move and around, maintain balance.
Feeling	Numerous sensors in its head, limbs and his sonars, enable it to perceive the environment and know its bearings.
Hearing and speaking	It interacts with humans in a completely natural manner through listening and speaking with 4 microphones and loudspeakers.
Seeing	Equipped with two cameras in high resolution, it can recognise shapes and objects.
Thinking	The robots can reproduce human behaviour.

Fig. 8 Information about Nao (Softbank 2017). *Note* It should be noted that these robots are for suggestions only. In case other cheaper robots are available in the market for sale with better function in Hong Kong, such as Zenbo later, we may adopt them instead

6 Smart Home Facilities Provided for the Elderly in the Community

In this project, we propose to liaise with NGOs which specialises in elderly services and a higher educated with disabilities organisation help us distribute the smart home devices to 100 elderly living in Kwai Tsing and Sham Shui Po District, especially on Lai King Estate, Cho Yiu Chuen, Shek Lei II Estate and Fu Cheong Estate who is suffering from chronic ill patients and their caregivers, especially on stroke and Mild Cognitive Impairment (MCI). Service users can self-approach or be referred from specialists in South Kwai Chung Social Service. These estates are selected mainly because there are many elderly in these estates than the others. The background information of the above estates is as follows:

	Lai King Estate	Cho Yiu Chuen	Fu Cheong Estate	Shek Lei II Estate
Population	11,200	16,439	16,017	23,700
Population of age > 60	4,380	3,951	4,022	3,024
Percentage of older persons (age > 60)	39	24	25.1	12.8
Number of households	4,200	2,533	5,900	8,300
Number of households (older persons who living alone and living with spouse only)	490	448	1,400	1,676
Percentage of households (older persons who living alone and living with spouse only)	11.7	17.7	23.7	20.2

Fig. 9 JACO assistive robotic arm (Kinova Robotics, 2017)

7 Conclusion

Smart home is one of the major features in smart city, smart home robots can act as companion to the elderly, smart wearable robots can help the disabled in feeding themselves, smart gadgets can save electricity and water. Understanding on smart home is limited among laymen in Hong Kong, elderly and disabled persons are scarce.

The proposed research incorporates three generations of smart home gadgets (a) smart home gadgets which are connected by network system; (b) smart gadgets which have certain level of artificial intelligence which can reply to our questions; (c) smart home robots which can move around the home to move around the home easily on behalf of the elderly and disabled persons.

Installing smart home gadgets at affordable price for 200 households with elderly/disabled persons to inform the elderly and disabled persons can help us study the impact of these tools in helping their daily lives. Whilst we can find many of the smart home laboratory in overseas universities such as University of West Florida, Iowa State University ISU Smart Home Lab, we cannot find any smart home laboratory designed for elderly and disabled persons. Academically speaking, the smart home gadgets install in smart home laboratory shall provide important information to university researchers to study people's behaviour and perceptions on smart home.

Acknowledgment This work described in this paper was fully supported by a grant from the Research Grants Council of the Hong Kong Special Administrative Region, China (Project No. UGC/IIDS15/E01/19).

References

Chan, M., Campo, E., Estève, D., & Fourniols, J.-Y. (2009). Smart homes—Current features and future perspectives. *Maturitas, 64*, 90–97.

Gaddam, A., Mukhopadhyay, S. C., & Gupta, G. S. (2010). Smart home for elderly using optimized number of wireless sensors. *Advances in Wireless Sensors and Sensor Networks, 64*, 307–328.

Gupta, S. (2015) For the disabled, smart homes are home sweet home. http://www.fortune.com/2015/02/01/disabled-smart-homes/.

Jarzebski, M. P., Elmqvist, T., Gasparatos, A., Fukushi, K., Eckersten, S., Haase, D., Goodness, J., Khoshkar, S., Saito, O., Takeuchi, K., Theorell, T., Dong, N., Kasuga, F., Watanabe, R., Sioen, G. B., Yokohari, M. & Pu, J. (2021). Ageing and population shrinking: implications for sustainability in the urban century. *npj Urban Sustainability, 1*, 17.

Li, R. Y. M. (2013). The usage of automation system in smart home to provide a sustainable indoor environment: A content analysis in web 1.0. *International Journal of Smart Home, 7*, 47–60.

Li, R. Y. M., Li, H. C. Y., Mak, C. K., Tang, T., & Beiqi,. (2016). Sustainable smart home and home automation: Big data analytics approach. *International Journal of Smart Home, 10*, 177–187.

Mutual Mobile. (2015). Smart home technology: Providing independence to the elderly while strengthening family ties. https://www.mutualmobile.com/resources/smart-home-technology-providing-independence-to-the-elderly-while-strengthening-family-ties.

Robson, S. M., Hansson, R. O., Abalos, A., & Booth, M. (2006). Successful aging: Criteria for aging well in the workplace. *Journal of Career Development, 33*, 156–177.

Sin, C. K. L., Fu, S. N., Tsang, C. S. H., Tsui, W. W. S., & Chan, F. H. W. (2015). Prevention in primary care is better than cure: The Hong Kong reference framework for preventive care for older adults—Translating evidence into practice. *Hong Kong Medical Journal, 21*, 353–359.

Stefanov, D. H., Bien, Z., & Bang, W. C. (2004). The smart house for older persons and persons with physical disabilities: Structure, technology arrangements, and perspectives. *IEEE Transactions on Neural Systems and Rehabitation Engineering, 12*, 228–250.

The Hong Kong Special Administrative Region of the People's Republic of China. (2021). The Chief Executive's 2021 Policy Address. https://www.policyaddress.gov.hk/2021/eng/p112.html.

Yigitcanlar, T., Corchado, J. M., Mehmood, R., Li, R. Y. M., Mossberger, K., & Desouza, K. (2021). Responsible urban innovation with local government artificial intelligence (AI): A conceptual framework and research agenda. *Journal of Open Innovation: Technology, Market, and Complexity, 7*, 71.

Chapter 7
Predicting Housing Price in Beijing Via Google and Microsoft AutoML

Dongming Chen and Rita Yi Man Li

Abstract For years, the hedonic regression model has dominated housing price research worldwide. However, the hedonic regression model suffered from the problem of over-simplification and heterogeneity. Machine learning has become a hot method in housing price prediction in recent years. The machine learning method in predicting housing prices is more accurate and precise than the traditional methods. This paper introduced three regression methods in housing price prediction: the traditional hedonic regression model, Google AutoML and Microsoft AutoML. It reviewed the factors that affected housing prices in literature and used the dataset of the housing price in Beijing in Kaggle to study the factors affected the housing price in Beijing. The results showed that Google AutoML had the best performance in predicting housing prices in Beijing. It had the highest R square (0.820) and the least RMSE and MAE. The average housing price in a community was the most important feature that impacted housing price prediction. Number of days open for sale and geographical location ranked the second and the third most important features in predicting the housing price.

Keywords Machine learning · Beijing · AutoML

1 Introduction

Housing price reflects the nature and the practical circumstances of selling in the open market (Pagourtzi et al., 2003). Housing can be used as a shelter and an investment tool. The accurate housing price prediction help us make investment decisions effectively for real estate developers and individual investors (Lam et al., 2008).

There are various determinants of housing price. Some scholars believed that the economic fundamentals such as income and interest rate are the main determinants

D. Chen
Department of Economics and Finance, Hong Kong Shue Yan University, North Point, Hong Kong

R. Y. M. Li (✉)
Sustainable Real Estate Research Center/Department of Economics and Finance, Hong Kong Shue Yan University, North Point, Hong Kong

R. T. M. Li et al. (eds.), *Current State of Art in Artificial Intelligence and Ubiquitous Cities*, https://doi.org/10.1007/978-981-19-0737-1_7

in housing price (Fang et al., 2016; Han et al., 2018). Some scholars addressed that housing characteristics such as the number of bedrooms, square meter and distance to the subway are the main determinants of housing price (Dai et al., 2016; Sun et al., 2016; Zhang et al., 2014). The hedonic pricing model is often used as a research method for estimating the effect of housing and environmental characteristics on housing pricing (Li et al., 2021). The hedonic pricing model is the utility-based regression model, which empathizes the preference of the housing characteristics for customers and is the most determining factor for housing price. However, the hedonic pricing model has long been criticized for being over-simplified and heteroscedastic (Stevenson, 2004; Fletcher et al., 2000). The significant drawbacks of the hedonic pricing model have provided an incentive for researchers to find a more accurate way to predict the housing price.

Among all methods in predicting housing prices, machine learning is one of the most popular methods. They are a kind of artificial intelligence technique for automating solutions to study and predicted complicated problems. Machine learning involves the process of training and testing data. With training and testing, machine learning can achieve the best prediction result for the housing price than the traditional regression method (Selim, 2009).

Beijing is a city with high population growth. The permanent population of Beijing has grown from 13.636 million to 21.536 million from 2000 to 2019, and the permanent migrant population has grown from 2.561 million to 7.456 million from 2000 to 2019 (Beijing statistical yearbook 2020), with a 57.93% growth of permanent population and a 191.14% growth of permanent migrant population. The increasing population has also increased the burden of housing supply in Beijing.

Kaggle is the competition platform of data analysis and modelling, dedicated to "making data science a sport". Data scientists and data science professionals crawled numerous datasets on this platform. So, this paper used the dataset "Housing price in Beijing" in Kaggle to run the model both in hedonic and machine learning models.

Google AutoML is a simple method for non-coding professionals to develop machine learning models. In Google AutoML's interface, we can import the dataset and see how much percentage explanatory of one variable to the independent variable is. This study utilised the following steps for developing machine learning in Google AutoML: 1. Imported the data, 2. Trained the data, 3. Tested the data and 4. Deployed the model.

Similarly, Microsoft AutoML is also the simple method to develop machine learning models and develop the predictability, performance and forecasting error for the model. However, Microsoft AutoML has one unique point compared to Google AutoML: Microsoft AutoML can run multiple algorithms simultaneously, and Microsoft AutoML can run 42 algorithms at one time.

2 Literature Review

Housing is a shelter and an investment tool that can generate a considerable income and cash flow for real estate developers and individual investors (Chen et al., 2017). The simple logic for real estate arbitrage is to buy low and sell high. By this method, investors can quickly gain the opportunity of appreciation and preservation. Therefore, housing prices are of great interest for real estate developers, potential and actual homeowners (Schulz & Werwatz, 2004). Beijing's housing price recorded a significant increase in Beijing's real housing price by 19.8% each year between 2006 to 2014 (Wu et al., 2012).

The rapid increase of housing prices in Beijing has attracted many scholars to investigate the reasons. Many scholars have estimated the various factors affecting the housing price in Beijing. Some scholars focused on investigating the influences of economic and population factors on the upsurge of the housing price in Beijing. For example, the increase of income per capita and large population inflow were the significant factors that drove the upsurge of housing prices in Beijing (Fang et al., 2016; Garriga et al., 2016). However, these research had room for improvement because they just focused on the role of consumption of housing neglected the role of investment in housing (Han et al., 2018).

The hedonic pricing model is a utility-based model, which empathizes the location and internal characteristics of a house are the primary determinant of the housing price (Rosen, 1974). The hedonic pricing model is a linear, or we say, intrinsically linear model (Montero, 2018). The hedonic pricing model includes four main forms:

1. Linear form;
2. Log-Linear form;
3. Linear-Log form;
4. Log-Log form.

In the previous paper, scholars summarized the main determinants of the housing price are: distance to the city centre (CBD), distance to subway stations, distance to nearest leisure facilites, number of bedrooms and building age using hedonic pricing model (Sun et al., 2016; Wang et al., 2012). They found that there exhibited a negative correlation between housing price and distance to the city centre, distance to subway stations and building age.

However, the hedonic pricing model was criticized for being overly simplified and had heteroscedasticity problem (Selim, 2009; Stevenson, 2004). To improve the accuracy of the prediction on housing price, some scholars improved in the hedonic pricing model by introducing spatial dynamic or smoothing data (Chen, 2017; Montero et al., 2018), and some of the scholars have started to use machine learning to make the housing price prediction.

The most common machine learning method in predicting the housing price is Artificial Neural Networks (ANN). Based on the previous studies in using ANN as a means of housing price prediction, ANN has been examined to have more excellent performance even though the sample size is relatively small (Lam et al., 2008). The

ANN performs a higher R square and lower Mean Standard Error (MSE) than the hedonic pricing model (Selim, 2009), which indicates that ANN is an effective way to predict housing price. Also, the ANN model outperformed the Support Vector Machine and ARIMA models (Abidoye et al., 2019).

Although no previous studies are using AutoML to predict housing prices in Beijing, it is hypothesized that Machine Learning methods have higher predictability and a lower prediction error than traditional hedonic pricing models.

Based on the study conducted by Park and Bae in 2015 (Park et al., 2015), this study consisted of five steps in constructing a machine learning model for housing price prediction:

1. Retrieve the dataset;
2. Divide the dataset into a training dataset and test dataset;
3. Choose the training parameters;
4. Use training dataset to train the model;
5. Test the model using the test dataset.

Sanjar et al. (2020) and Truong et al. (2020) used Kaggle as the database to develop the machine learning model for predicting the housing price. This research follows Sanjar et al. and Truong et al.'s way of obtaining the dataset.

By reviewing the literature, two research questions are included:

1. What are the major factors that affect housing prices in Beijing?
2. Do Machine learning methods have a better prediction performance compared to traditional hedonic models?

This paper retrieved the Beijing housing data from Kaggle and used Google AutoML and Microsoft AutoML to construct a machine learning model in linear regression. This study compared the machine learning models' performance by studying R square, Root Mean Standard Error (RMSE) and Mean Average Error (MAE) and then compared them with the hedonic pricing model.

3 Data Source and Variables

This study adopted data from Kaggle, a competition platform for data modelling and analysis. In Kaggle, the datasets are either issued by the company or crawled by data science professionals. Kaggle has organized a data competition event with NASA, Wikipedia and Deloitte.

The success of Kaggle has also attracted interest from Venture Capital and Private Equity. In November 2011, Kaggle generated 11 million USD from Index Venture, Khosla and other VCs.

Therefore, the dataset in Kaggle is reliable because it has many data professionals in this platform, which is a promise for high-quality datasets and algorithms.

The dataset "housing price in Beijing" was fetched from Lianjia.com, one of the most significant transaction platforms for real estate in Mainland China. The dataset

Table 1 Variables in the dataset

Variable	Meaning
Price	Price/m
Lng	Longitude of the housing
Lat	Latitude of the housing
DOM	The number of days in the market
Livingroom	Number of livingrooms
Kitchen	Number of kitchens
Bathroom	Number of bathrooms
Floor	Floor height of the housing
Buildingstructure	Structure of the housing
Ladderratio	The proportion between number of residents on the same floor and the number of elevator
Elevator	Whether the housing have elevator or not
Subway	Whether the housing near subway station or not
Community Average	Average housing price in that community

contains 25 unique variables and 318,852 housing data in Beijing arranged from 2011 to 2017. This amount of data is sufficient for the machine learning model. 13 important variables were included for the analysis. The variables and the meanings are listed in Table 1.

Price is the price (in RMB) per square meter. Lng is the longitude of the housing. It measures the geographical position of the west and east. The greater the Lng is, the housing is facing east. Lat is the latitude of the housing. It measures the geographical position of the north and south. The greater the Lat is, the housing is closer to face north. DOM indicates the number of days in the open market. The larger the DOM, the longer it has stayed in the market. Living room, Kitchen, and Bathroom refer to the number of living rooms, kitchens, and bathrooms. The floor measures the floor height of the housing. Building structure includes mixed, brick and wood, steel, steel-concrete composite, brick and concrete, and unknown.

Ladder ratio measures how many elevators per resident are on the same floor. Elevator and subway refer to whether the housing has an elevator or subway, and the community average is the average price.

4 Model Selection

We selected the linear–linear form hedonic pricing model for the traditional hedonic pricing model since some of the variables contained zero, we could not take the log in Eviews. Price/square meter was the dependent variable in this model, while others were independent variables.

For each variable, we looked at the different t-statistics and reviewed whether the variable was significant. A table that contained the variable names, coefficients and p-values for each variable were listed. At last, we displayed the forecast performance for the traditional hedonic model.

For the Google AutoML, we chose the "price" as the dependent variable and then removed all the empty data before running the machine learning model. We followed the original setting of the Google AutoML and ran the model.

Although both Google AutoML and Microsoft AutoML perform machine learning automatically, Microsoft AutoML has an entirely different interface than Google AutoML. They have a completely different operation.

5 Research Result

Firstly, we will show the regression result of the traditional hedonic model. Table 2 is the result for the hedonic model. The signal "*" measures the different levels of significance for the variables, and the variable with more "*" indicates it is more significant.

Table 2 Coefficient of hedonic pricing model

Variables	Coefficient
C	783,905.2***
Lng	−6,291.410***
Lat	−1,186.931***
DOM	162.8308***
Livingroom	−1410.269***
Kitchen	−113.6382
Bathroom	−9.789660***
Floor	−74.92573***
Building structure	29.06400
Ladderratio	0.001635
Elevator	13.97693
Community average	0.631207***

*-Significant at 10% level;
**-Significant at 5% level;
***-Significant at 1% level

In Table 2, we can see that the number of kitchens, building structure, ladder ratio, and elevator are not significant factors affecting Beijing's housing price. The other variables are all significant at the 1% level, which is the best performance among our chosen criteria.

Next, we use the "forecast" function in Eviews and see the performance of the traditional hedonic model on the prediction of housing prices in Beijing. The forecast performance for the hedonic pricing model is shown in Figs. 1 and 2. We summed up all the prediction results in Table 3.

Figures 1 and 2 show that the hedonic pricing model has an R square of 0.536126, RMSE of 14,814.82 and an MAE of 10,450.94. In this regression model, the performance is not very good because the data set has more than 310,000, and the number of variables is 13, but R squared is only about 3% more than 50%. In general, the higher the number of data and variables, the higher the R squared is, the more likely reason is that hedonistic regression models are suspected of false regressions.

R-squared	0.536126	Mean dependent var	43477.30
Adjusted R-squared	0.536108	S.D. dependent var	21751.89
S.E. of regression	14815.12	Akaike info criterion	22.04472
Sum squared resid	7.00E+13	Schwarz criterion	22.04516
Log likelihood	-3514478.	Hannan-Quinn criter.	22.04485
F-statistic	30708.29	Durbin-Watson stat	0.732891
Prob(F-statistic)	0.000000		

Fig. 1 R-square of hedonic pricing model

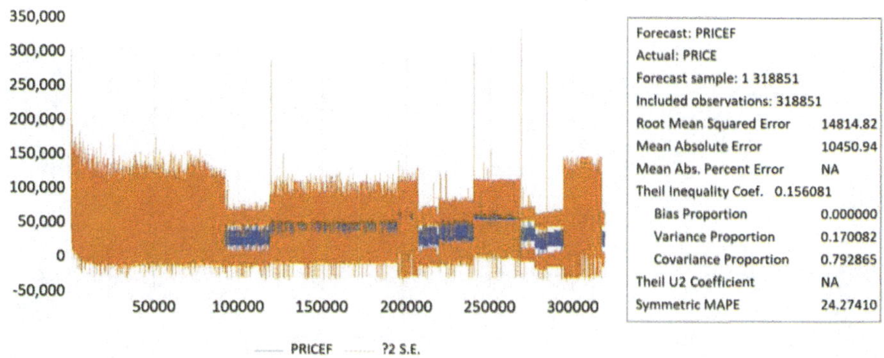

Fig. 2 The forecasting performance of the hedonic pricing model

Table 3 .

Methods	R square	RMSE	MAE
Hedonic pricing model	0.536126	14,814.82	10,450.94
Google AutoML	0.820	9,347.760	6,557.418
Microsoft AutoML (Voting ensemble)	0.792	9,952.922	7,130.104

Among all the algorithms that Microsoft AutoML ran, the algorithm with the best performance is the Voting Ensemble algorithm. It had an R square of 0.792, an RMSE 9,952.922 and an MAE of 7,130.104.

Comparatively speaking, the hedonic pricing model is underperformed than Google AutoML and Microsoft AutoML in predictability and error. It is believed that the procedure of "training with the data" in Machine learning effectively improves the predictability of the regression model.

Among these three methods, Google AutoML has the best performance in predicting housing prices in Beijing because of the highest R square (0.820) and least RMSE and MAE. Google AutoML has also calculated the importance of one variable to predict housing prices in Beijing. The feature importance for each variable is shown in Fig. 3.

Figure 3 indicates that the most crucial feature in predicting the housing price in Beijing was the average housing price in a community. The second important feature was the days on the market. Geography position was the third essential feature in predicting the housing price in Beijing.

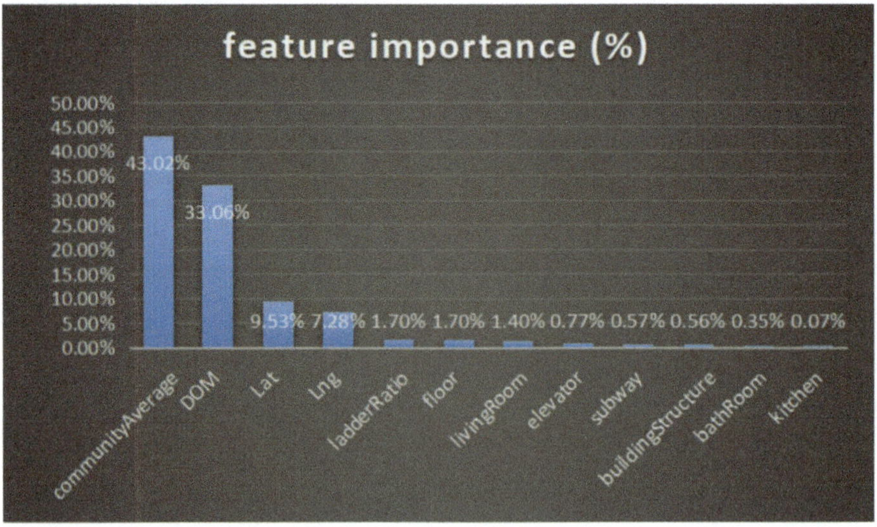

Fig. 3 Feature importance of variables

However, according to the prediction result from Google AutoML, the housing characteristics such as the building structure, number of living rooms, kitchens, bathrooms were not the most factors in predicting the housing price in Beijing. On the other hand, whether the housing is near a subway station or not and whether the housing has an elevator are not the factors to predict housing prices in Beijing.

The possible reason for this result was that expensive residential areas tend to have expensive buildings, and cheap residential areas tended to have cheap buildings. This also made the community average a significant factor in the prediction of machine learning, so in the machine learning model for predicting the housing price of Beijing, the community average accounted for 43.02% of the overall prediction variable. In addition, housing prices in Beijing can also be predicted from how long a building has been left on the open market, possibly because on online real estate trading platforms, buyers judge the popularity of a house by how many days it has been on the platform. If a house has been on the Internet for a long time and had not been sold, it might be that the house was not popular or some hidden factors. Another essential factor predicting housing prices in Beijing is the geographic factor of housing. There is an old saying in Beijing: "people who live in the east city are rich, people who live in the western city are honourable, people who live in the northern city are poor, and people who live in the southern city are degrading." Although he is just a folk saying and academically groundless, it also illustrates the difference in income and living standards of people living in different areas in Beijing.

6 Conclusions and Implications

In this study, we compared the predictability of the traditional hedonic pricing model with machine learning developed from Google AutoML and Microsoft AutoML. We found that machine learning had a better performance in housing price prediction in Beijing than traditional hedonic pricing model. Google AutoML had the best performance in predicting housing prices in Beijing among all models. It achieved the highest R square, the least RMSE and MAE. The most important factors in predicting the housing price in Beijing was the average housing price in a community. Number of days open for sale in the market ranked second. Geographical location was the third most important feature in predicting the housing price in Beijing. In the future, we might expand the study to other cities in China, test the predicting performances by using other machine learning and forecasting approaches.

Acknowledgment This work described in this paper was fully supported by a grant from the Research Grants Council of the Hong Kong Special Administrative Region, China (Project No. UGC/IIDS15/E01/19).

References

Abidoye, R., Chan, A., Abidoye, F., & Oshodi, O. (2019). Predicting property price index using artificial intelligence techniques. *International Journal of Housing Markets and Analysis, 12*(6), 1072–1092. https://doi.org/10.1108/ijhma-11-2018-0095

Chen, J., Ong, C., Zheng, L., & Hsu, S. (2017). Forecasting spatial dynamics of the housing market using support vector machine. *International Journal of Strategic Property Management, 21*(3), 273–283. https://doi.org/10.3846/1648715x.2016.1259190

Dai, X., Bai, X., & Xu, M. (2016). The influence of Beijing rail transfer stations on surrounding housing prices. *Habitat International, 55*, 79–88. https://doi.org/10.1016/j.habitatint.2016.02.008

Fang, H., Gu, Q., Xiong, W., & Zhou, L. A. (2016). Demystifying the Chinese housing boom. *NBER Macroeconomics Annual, 30*(1), 105–166.

Fletcher, M., Gallimore, P., & Mangan, J. (2000). Heteroscedasticity in hedonic house price models. *Journal of Property Research, 17*(2), 93–108.

Garriga, C., Tang, Y., & Wang, P. (2016). Rural-urban migration, structural transformation, and housing markets in China. Working Paper Series #2014–028c, Federal Reserve Bank of St. Louis.

Han, B., Han, L., & Zhu, G. (2015). Housing price and fundamentals in a transition economy: the case of Beijing market. *SSRN Electronic Journal*. https://doi.org/10.2139/ssrn.2664817

Han, B., Han, L., & Zhu, G. (2018). Housing price and fundamentals in a transition economy: The case of the Beijing market. *International Economic Review, 59*(3), 1653–1677

Lam, K., Yu, C., & Lam, K. (2008). An artificial neural network and entropy model for residential property price forecasting in Hong Kong. *Journal of Property Research, 25*(4), 321–342. https://doi.org/10.1080/09599910902837051

Li, N., Li, R. Y. M., & Pu, R. (2021). What is in a name? A modern interpretation from housing price in Hong Kong, Pacific Rim Property Research Journal, 27(1), pp. 55–74

Montero, J., Mínguez, R., & Fernández-Avilés, G. (2018). Housing price prediction: Parametric versus semi-parametric spatial hedonic models. *Journal of Geographical Systems, 20*(1), 27–55. https://doi.org/10.1007/s10109-017-0257-y

Park, B., & Bae, J. (2015). Using machine learning algorithms for housing price prediction: The case of Fairfax County, Virginia housing data. *Expert Systems with Applications, 42*(6), 2928–2934. https://doi.org/10.1016/j.eswa.2014.11.040

Pagourtzi, E., Assimakopoulos, V., Hatzichristos, T., & French, N. (2003). Real estate appraisal: a review of valuation methods. Journal of Property Investment & Finance.

Rosen, S. (1974). Hedonic prices and implicit markets: Product differentiation in pure competition. *Journal of Political Economy, 82*(1), 34–55. https://doi.org/10.1086/260169

Sanjar, K., Bekhzod, O., Kim, J., Paul, A., & Kim, J. (2020). Missing data imputation for geolocation-based price prediction using KNN–MCF method. *ISPRS International Journal of Geo-Information, 9*(4), 227. https://doi.org/10.3390/ijgi9040227

Schulz, R., & Werwatz, A. (2004). A state space model for Berlin house prices: Estimation and economic interpretation. *Journal of Real Estate Finance and Economics, 28*(1), 37–57.

Selim, H. (2009). Determinants of house prices in Turkey: Hedonic regression versus artificial neural network. *Expert Systems with Applications, 36*(2), 2843–2852. https://doi.org/10.1016/j.eswa.2008.01.044

Sun, H., Wang, Y., & Li, Q. (2016). The impact of subway lines on residential property values in Tianjin: An empirical study based on Hedonic pricing model. *Discrete Dynamics in Nature and Society, 2016*, 1–10. https://doi.org/10.1155/2016/1478413

Stevenson, S. (2004). New empirical evidence on heteroscedasticity in hedonic housing models. *Journal of Housing Economics, 13*(2), 136–153. https://doi.org/10.1016/j.jhe.2004.04.004

Truong, Q., Nguyen, M., Dang, H., & Mei, B. (2020). Housing price prediction via improved machine learning techniques. *Procedia Computer Science, 174*, 433–442. https://doi.org/10.1016/j.procs.2020.06.111

Wu, J., Gyourko, J., & Deng, Y. (2012) Evaluating conditions in major Chinese housing markets. *Regional Science and Urban Economics, 42*, 531–543.

Zhang, M., Meng, X., Wang, L., & Xu, T. (2014). Transit development shaping urbanization: Evidence from the housing market in Beijing. *Habitat International, 44*, 545–554. https://doi.org/10.1016/j.habitatint.2014.10.0

Chapter 8
AI Object Detection, Holographic Hybrid Reality and Haemodynamic Response to Construction Site Safety Risks

Rita Yi Man Li, Kwong Wing Chau, and Daniel Chi wing Ho

Abstract Construction practitioners make decisions about safety risks that can be subjective and prone to error. The trained computer object detection provides a standardised method to deal with this issue, but it typically relies on many photos to train one object, which is costly and time-consuming. This project proposes a new algorithm to train a computer to identify construction safety risks with fewer photos. In addition, holographic hybrid reality will be developed for safety training in the construction industry; Mercedes-Benz has used a similar approach to inform employees about collision zones. We will use the trained images to develop HoloLens hybrid reality to share construction site safety knowledge via wearable HoloLens for on-site safety risk detection. Lastly, although decision-making in various areas has been studied using neuroscience, how an individual's brain makes decisions when different construction safety risks are perceived and the impact of holographic safety training on brain reaction and activities remains unknown. These issues will be studied via haemodynamic response and neuroimaging. In this research agenda, we plan to construct a photo library of 10,000 high-quality photos of various construction risks with different shading, size, and orientation by collecting Creative Commons construction photos and turning existing online Creative Commons videos into photos. To achieve this goal, we will input the online common creative photos into our image-based CAPTCHA system (similar to Google's ReCaptcha). Each group will include 16 photos to allow construction practitioners to identify and click on those that include a safety risk. The identified photos with safety risks will be saved, and specific categories of safety risks will be uploaded to social media and sent via email. The trained figures will be deployed to the HoloLens in our laboratory. About 20 safety experts, including safety managers and trainers, will be invited to use the

R. Y. M. Li (✉)
Sustainable Real Estate Research Center/Department of Economics and Finance, Hong Kong Shue Yan University, North Point, Hong Kong

K. W. Chau
Department of Real Estate and Construction, The University of Hong Kong, Pok Fu Lam, Hong Kong

D. C. Ho
The Faculty of Design and Environment, Technological and Higher, Education Institute of Hong Kong, Chai Wan, Hong Kong

R. T. M. Li et al. (eds.), *Current State of Art in Artificial Intelligence and Ubiquitous Cities*, https://doi.org/10.1007/978-981-19-0737-1_8

HoloLens for detecting hazards on-site and provide comments for improvement. Holographic hybrid reality will be built with Unity C# and HoloToolkit. The object detection results obtained will also be used so the research participants will see the real scene with not only hazards labelled by AI, but also some high-risk elements that cannot be included in ordinary safety training, such as open taps with water running into the ground and blasting. Four holographic hybrid reality training scenarios will be generated: a general construction site, and three specially designed scenarios for refurbishment, new building, and road construction settings. In the last stage, we will use functional near-infrared spectroscopy (fNIRS) to study construction practitioners' brain responses when they see and identify various kinds of hazards. The first group will be exposed to holographic hybrid reality with some safety risks on the first day, and they will be asked on the tenth day to identify the safety risks. The control group will receive no safety training but will be asked to identify risks. All research participants will be monitored by fNIRS when they attempt to identify the safety risks. Haemodynamic response and neuroimaging tests will be used to study the effectiveness of the safety training.

Keywords Object detection · Construction safety · Functional near-infrared spectroscopy (fNIRS) · Hong Kong

1 Introduction

Project safety risk assessment is important for efficient project management because accidents frequently occur on construction sites. Moreover, occupational injuries can be as high as 15% of the total project costs (Koulinas et al., 2019). One study suggested that local hazards related to fixed installations and usability of materials and equipment account for 40% and 42% of accidents on construction sites, respectively (Winge et al., 2019). Many safety risks are visible to the human eye and can be prevented by proper risk assessment. However, inattentional blindness may occur, in which people fail to notice stimuli in front of them without experiencing a vision defect or deficit (Liao & Chiang, 2016). Moreover, human judgments about safety risks are often subjective and prone to error due to varying degrees of knowledge and experience with hazardous conditions and safety (Pandit et al., 2019).

Consequently, Chan et al. (2018) advocated using AI tools to improve the quality of task-specific safety risk assessments for infrastructure projects by facilitating the retrieval of knowledge and experience. However, they did not utilise AI object detection to identify systematic safety risks.

AI object detection using objective criteria increases the likelihood of identifying on-site safety risks, thus can be a valuable tool for risk assessment. Nevertheless, traditional object detection relies on numerous images for training, which requires an extensive photo database. Moreover, similar to the human learning process, the more items a computer must learn, the easier it is to confuse them. Indeed, many more are required for training purposes for any additional item to be detected.

We propose to involve the generation of a photo library with 10,000 photos regarding safety risks, such as uneven floors and electrical wires running across floors (i.e., trip hazards), which will be used to train the computer to identify on-site risks. This AI object detection will help construction practitioners to identify hazards in a more standardised way and will require fewer photos than previous methods due to the use of a new algorithm. Besides, practical safety training usually involves an emotional connection to the subject so that a learner-centric andragogical knowledge-sharing approach may improve safety knowledge's cognitive change (Loosemore & Malouf, 2019). For this reason, the proposed study will involve holographic safety training with wearable HoloLens. As the research participants learn, the cognitive load should decrease, as reflected in lower oxyhaemoglobin levels. Thus, a haemodynamic response will be used in this study to investigate training effectiveness. In this research, the aims are fourth folds:

1. To construct a comprehensive construction site hazard photo library to train a computer to identify on-site safety risks.
2. To build a holographic hybrid reality for safety training and safety risk identification.
3. To study brain decision making and safety risk identification via haemodynamic response and neuroimaging.
4. To investigate the effectiveness of holographic hybrid reality safety training and knowledge retention in memory.

In the following sections, Sect. 2 will be the literature review, Sect. 3 will be the research methods and Sect. 4 will be the conclusion.

2 Literature Review

2.1 Computer Vision

Traditional semantic segmentation and classification only allow the identification of a single object, whereas AI object recognition, which allows the identification of multiple objects, has recently come to the fore. The most common application is facial recognition. For example, some restaurants in China allow customers to authorise digital payment by facial scan (Larson, 2018). Although object recognition can distinguish one object from another, it cannot show the coordinates, so it is impossible to know where the objects are located. Object detection is helpful because it allows the detection of multiple objects and their locations.

In the construction industry, Son et al. (2019) used a classification approach to detect the presence of workers by Faster-CNN with bounding boxes. Nonetheless, this classification approach can only detect one object, limiting practical applications for hazard identification because there are many different types of on-site hazards. Moreover, their study did not show the probability level of each hazard, so there is no standardised benchmark for the existence of a hazard. While Luo et al. (2018) utilised CNN to detect the construction workers' activities, they did not study the construction safety risks. Thus, there is a need to develop object detection for many construction site safety risks to suit real-life situations better.

2.2 Holographic Hybrid Reality for Knowledge Sharing

Previous research evidenced that visual communication enhanced knowledge sharing. For example, infographics improve construction safety knowledge transfer (Lingard et al., 2018). Golizadeh et al. (2018) suggested that digital planning to address construction accident causes is a potential field for future research.

In contrast to virtual reality, holographic hybrid reality allows users to interact with the real environment via overlaid virtual objects (Incekara et al., 2018) without experiencing motion sickness, which is common in virtual reality (Matthews, 2018). Spatial mapping provides a detailed representation of real-world surfaces around the HoloLens, allowing a convincing hybrid reality experience. Spatial sound allows developers to place sounds in a 3D sphere around the user convincingly. Holograms, objects made of light and sound, can respond to gaze, air, tap and voice commands, interacting with real-world surfaces around users so that they appear to be real objects; none of these features is included in traditional augmented reality (Li 2018b).

A previous study used holographic hybrid reality to generate magnetic resonance imaging-based three-dimensional holograms of patients' tumours and project them on 25 patients' heads with the HoloLens (Incekara et al., 2018). Pearson's HoloChemistry and HoloMath have been available for purchase since September 2018. In construction research, a quasi-experiment compared the performance of 18 electrical construction personnel who were tasked with building similar conduit assemblies using traditional paper and hybrid reality with head-mounted HoloLens at the design stage. It was found that the latter improved productivity by reducing the time needed to understand the design (Chalhoub & Ayer, 2018), as high engagement training methods require active worker participation, resulting in better training results (Namian et al., 2016). Nevertheless, holographic hybrid reality with AI object detection and wearable HoloLens has never been used to aid construction workers to identify construction safety risks or knowledge sharing.

We hypothesise that (H1) holographic hybrid reality with AI object detection will help construction practitioners to identify safety risks on sites, and (H2) holographic hybrid reality can be an effective means for safety knowledge retention due to its high engagement in nature and special holographic features.

2.3 Haemodynamic Responses to Decision-Making

Classical economic theory following Bentham and Mill postulates that decision-makers are entirely informed, highly sensitive to differences between alternatives, and able to make rational choices following utility maximisation. However, this view has been successfully challenged by a plethora of experiments on decision-making (Burnett, 2008). The diffusion model, one most influential models in Neuroeconomics, advocates that the brain monitors the difference among alternatives. When the difference reaches a threshold, the firing of neurons for decision making attains the threshold and confirms the choice (Li et al., 2019, in press).

Cognitive workload measures the effort associated with a task (Curtin & Ayaz, 2018), so when a brain area is activated to execute a task, the metabolic demand for oxygen and glucose increases, leading to an oversupply in regional cerebral blood flow (CBF) to meet this demand. Functional hyperaemia, an increase in CBF in response to an increase in neuronal activity, is mediated by neurovascular coupling mechanisms, such as changes in vasoactive metabolites and capillary diameter. Oxygen reaches the activated brain region more quickly than it is consumed. Thus an oversupply in regional CBF leads to an increase in oxyhaemoglobin and a decline in deoxygenated haemoglobin concentrations, known as the haemodynamic response. These can be estimated by changes in light attenuation measured by functional near-infrared spectroscopy (fNIRS) (Pinti et al., 2019, forthcoming).

As individuals become more competent at tasks, they adopt less strenuous strategies. Therefore this increased expertise results in reduced cognitive demands and brain activity (Milton et al., 2007). For example, Curtin and Ayaz (2018) showed that as the participants progressed from inexperienced to more experienced unmanned aerial vehicle operators, oxyhaemoglobin gradually decreased, mirroring the improvement in their performance.

Thus, we hypothesise that (H2a) there will be a decrease in the level of increase in oxyhaemoglobin and the level of decline in deoxygenated haemoglobin after holographic training compared to baseline condition, indicating that trainees are more competent in safety risks identification.

2.3.1 Regions of the Brain That Are Activated

According to de Winkel et al. (2017), the prefrontal cortex plays a central role in decision-making under uncertainty and risk. Moreover, more activity is observed within the left dorsolateral prefrontal cortex during an easier decision than during a difficult one (Heekeren et al., 2004). Jahani et al. (2017) adopted fNIRS to investigate the encoding of novel face-name pairs compared to baseline and showed that repeating face-name pairs resulted in significant activation of the left dorsolateral prefrontal cortex.

In addition, a previous study showed that people's cognitive demand is different when they perceive different things. For example, blood oxygen levels in the brain are higher when viewing faces than houses, illustrating that we have different responses to different things (Heekeren et al., 2004). Megías et al. (2018) revealed that a rise in the blood-oxygenation level is related to higher levels of perceived driving risk, leading to a rise in neural activity in the insula, inferior parietal lobule, cingulate gyrus, thalamus, fusiform gyrus and cerebellum etc. Higher activation in the cerebellum, thalamus, and putamen occurred only with high risk, while the supplementary motor area showed increased activation in response to medium and high risk.

In the construction industry, Jebelli et al. (2018) proposed to use an electroencephalogram signal-processing framework, a tool widely used in neuroscience, to examine whether eight workers were active or not. Another team of three construction researchers developed a wearable electroencephalographic safety helmet which enabled the collection of power spectral densities in the brain (Chen et al., 2016).

Despite many studies on the haemodynamic response in various activities, there is a lack of research regarding brain activation due to construction safety risks by fNIRS. Accordingly, we hypothesise that (H2b) after holographic hybrid reality safety training, hazard identification will be easier; more activity will be evident in the left dorsolateral prefrontal cortex in neuroimaging, and (H3) the areas activated by perceived high, medium, and low construction hazards will be similar to those activated by driving risks.

2.4 Episodic Memory of Knowledge Sharing

There is much research based on memory. For example, Li et al. (2019, in press) suggested that knowledge is a mixture of values, constructing new experiences based on the earlier elaboration in memory. Interpretation of how things work is an elaboration of the experiences in memory. Li (2018a) reported that our research assistants have a sense of falling, even after a few months of fall from height virtual reality training. Fear intrudes learners to have a deep memory, leading to better safety awareness and prevention. Lai et al. (2019) suggested that corporate memory is an institution that affects decision making in land reclamation.

Prior research revealed a strong link between episodic memory and episodic prospection. Indeed, the past holds valuable experiences which affect future choices and decisions (Prabhakar & Hudson, 2019). Thus, if the safety knowledge acquired from training is deep inside our memory, we have easier, faster, and more accurate safety decision making. Moreover, labelling encourages attention to featured information, strengthening our memory by encoding features that are relevant to the label (Kelly & Heit, 2017).

In an experiment with 12 volunteers watching videos of an expert stone toolmaker making axes, one group heard the craftsman's instructions, and the other watched without audio. The participants then made the tools themselves. The research showed that memory worked better when more than one sense was engaged, which prompted one museum to put jars of spices at the entrance to an Asian art exhibition (Curry, 2017). While it is possible to observe a disconnection between memory in the past and envision a similar future among children (Prabhakar & Hudson, 2019), such disconnection has not been studied in construction safety knowledge sharing.

Flashbulb memory implies a special memory mechanism triggered by an event exceeding critical levels of surprise, consequently creating a permanent record of the contents of awareness (Kryklywy et al., 2013). Limited level of fear-related stimuli acts as reinforcers, activating the amygdala and enhancing episodic memory. Dysfunctional emotional memory processes include psychiatric disorders such as post-traumatic stress disorder (Loos et al., 2019) and episodic memory deficits are related to unhealthy lifestyle such as uncontrolled eating among the Purdue University students in a 10-day study on a diet (Martin et al., 2018), alcohol consumption, Social engagement problem and poor sleep among the older people. Nevertheless, all these impacts on head-mounted HoloLens safety training is unknown.

We hypothesise that: (H4) episodic memory of holographic safety training is strongly linked with episodic prospection; (H5) labelling and more than one sense engagement enhances knowledge retention; (H6) a disconnection between memory in the past and envisioning the similar exists, i.e. people remember the holographic training but cannot identify the safety risks; (H7) holographic hybrid reality training triggers flashbulb memory. Thus, better knowledge retention; (H8) limited level of fear-related enhances episodic memory of the safety knowledge shared; (H9) unhealthy lifestyle may be attributed to episodic memory deficits.

3 Previous Research Methods

3.1 Computer Vision for Object Detections

Zhong et al. (2018) validated a position-sensitive balancing framework for 10-class geospatial object detection based on remote sensing imagery and a convolutional network to solve the dilemma between translation-invariance in classification and translation-variation in object detection. Wang et al. (2018) used a Faster-CNN object detection method to detect dairy goats by crucial frame extraction, foreground segmentation, and region proposal with a 92.49% precision rate. Fang et al. (2018) pioneered the Faster R-CNN method to detect construction workers who were not wearing a hat by using more than 100,000 construction worker images. With the newly developed algorithm, Custom Vision required only 50 images to begin object detection training, so time can be saved in generating a database of safety risk photos for training, which allows the system to be trained to identify many more hazard objects.

Li (2018b) utilised 25 high-quality photos for training a system to recognise electrical wires and holes on sites with a 99.8% accuracy rate via Custom Vision. However, no research utilised custom vision object detection algorithm to detect many different types of safety risks on sites, build a holographic hybrid reality for safety inspection and training or construction risks decision-making process via a neuroscience approach (i.e., neuroimaging and haemodynamic response).

3.1.1 Custom Vision Training

First, we will conduct a literature review of articles indexed in Web of Science and Scopus, the two most authoritative academic journal databases in academia, regarding visible safety risk objects on sites and the factors that affect the effectiveness of digital construction sites safety knowledge sharing.

We will include safety risk photos for new building construction sites, building refurbishments and road construction because (1) these construction activities are the most common in Hong Kong and (2) these are the only types of projects which have a readily available relevant Unity asset in the asset store for stage 2. Although forty hazard photos with more than one hazard per photo were collected for our previous studies, training the computer in safety risk identification with high accuracy rates requires at least 50 *high-quality* photos with a combination of different colours, sizes, lighting conditions, and shading for *one* salient safety risk object on a clean background. Moreover, as the number of objects increases, more high-quality photos are required for training purposes.

Thus, in addition to the pictures collected in our previous research projects, we will download all Creative Commons photos from Flickr and Pexels and draw on construction industry Creative Commons videos via Apowersoft Free Online Video Downloader. The downloaded videos will be converted to photos using the Free Video to JPG Converter. We will also send invitations to 319 construction companies in the Builders' Directory, construction and property management companies with long-term collaboration relationships and relevant online social media groups to obtain on-site safety risks photos for the abovementioned construction project types. Visible safety risks of green features in these projects will also be requested as they are more common, but there is a lack of study in this aspect.

We will include all the photos in our Image Captcha system, which has already been built (similar to Google Image reCAPTCHA, which determines whether a user is a human or a robot by asking them to identify some photos on the screen; see the simple version in GitHub built at https://github.com/siyanhu/simpcaptch). Like Kwon and Cha (2017), we will include typical questions such as 'Please select all the photos with ______.' (This blank will be filled with 'safety risks' at the beginning of the test to help us retain photos with safety risks. We will then use the specific names of safety risks, such as an electrical wire leading to a trip and fall accident.)

The main idea is to feed unlabelled images into the system and place them randomly as illustrated. The research participants will not be aware that they are performing image-labelling tasks, and they will be requested to hand-pick all images that fit the question. If unlabelled images are chosen in successful responses, we can obtain reliable tags automatically (Kwon & Cha, 2017). As previous researchers, including Google, utilised image CAPTCHA to obtain free and motivated image-labelling objects, we consider this a viable solution to identifying photos with safety risks from construction practitioners. We will post the image CAPTCHA to online groups on Facebook, LinkedIn, and email the Captcha to all construction practitioners in the Builders' Directory, inviting them to select and tag the safety risks shown in the figures.

Microsoft Custom Vision will then be applied to train the computer regarding 50 visually observable safety risks that are commonly found on sites. The Custom Vision Cloud Service is well-suited for object detection problems because it can evaluate images based on limited training, and images can be retagged when the computer makes the wrong interpretation of the safety risk. As the AI may be confused by too many safety risks, we expect that more than 50 quality photos will be needed, so we will include 10,000 quality photos in the first stage for training and testing because that is the limit of the Custom Vision S0 plan at present.

3.2 *Holographic Hybrid Reality*

According to Microsoft Custom Vision Documentation, we will deploy the trained results in stage 1 to HoloLens for safety risk detection on sites. A label tag will appear in the HoloLens for each of the actual safety risks detected with the probability of the risk level to aid construction practitioners in risk identification. We will invite about 20 safety officers and construction practitioners to use it on-site to validate and improve the application in detecting the actual safety risks.

Holographic hybrid reality training will be built for general construction sites, new buildings, refurbishment for existing buildings and road construction. By using these assets, we will develop the holographic hybrid reality using Unity with C# on Windows 10 Pro, Hyper V, and HoloToolkit. We will also use the Keyword Recognizer class to implement HoloLens speech commands. Although it is impossible to set a real fire or let a water tap run freely to show the danger in any real-world safety training scenario, the Unity and HoloToolkit allow participants to experience these potentially dangerous conditions.

Microsoft Human Response Transfer Function includes spatial sound to make the HoloLens experience more immersive. We will deploy the scenarios on top of the trained AI objects to a HoloLens emulator for debugging and testing, send it to the remote machine, then deploy it to HoloLens (Li, 2019) and the virtual elements simultaneously. We will invite safety officers and construction practitioners to use it on-site to validate and improve the holographic hybrid reality application's usefulness in detecting real safety risks and safety training.

3.3 *Haemodynamic Response and Neuroimaging*

Functional neuroimaging studies based on electroencephalography, positron emission tomography, functional magnetic resonance imaging, and fNIRS have been performed to study neurally correlated translation. fNIRS provides better spatial resolution than electroencephalography, and non-invasive mobile fNIRS has significant advantages in the study of cognitive neuroscience in daily settings, as it has very few body constraints (Ren et al., 2018) and a higher temporal resolution, allowing for rapid data acquisition up to 100 Hz, as compared to 1 Hz or less provided by fMRI. Furthermore, fNIRS provides real-time haemodynamic response recording, contributing higher-resolution information for real-time series analyses (Schaeffer et al., 2014). Compared to traditional magnetic resonance imaging, fNIRS is cheaper and safer than electroencephalography, as the latter may induce a seizure among some individuals (Li et al. 2019, in press).

To study the effectiveness of utilising holographic hybrid reality for safety training and the brain response to safety risks in a maximum of 48 research participants (due to the restriction of the software DOE) (Minitab, 2019). One group of research participants will receive safety risk identification training with HoloLens, and the control group will not receive any training. They will be recruited through the Safety group in Telegrams, LinkedIn, Facebook, and emails to all companies listed in the Hong Kong Construction Associations and Builders Directory. The participants should not have Alzheimer's disease, hypothyroidism, depression, sleep apnoea or celiac disease, leading to poor memory.

As HoloLens is sold mainly in the United States and Canada, we expect that no participants will have had HoloLens experience, so we will give them a quick tutorial on how to interact in the hybrid-reality environment by voice, gazing and tapping. Due to its intuitive nature, familiarising oneself with it is quick and easy (Raptis et al., 2018). Previous research showed that visual intelligence has a close relationship with IQ, so we will adopt the Wechsler Adult Intelligence test (WAIS-IV), which is considered the best IQ assessment for adults on the first day (Sellers et al., 2015).

Mobile fNIRS, which allows research participants to move around freely, will record the haemodynamic response when the participants make decisions and identify the safety risks on the first day we meet the experimental group for holographic hybrid reality training and control group. The traditional 10–20 system covered with a black cloth will be used to suppress surrounding light (Richter et al., 2018). Hair will be parted manually under the optodes to enhance signal detection (Fishburn et al., 2019). Spatial pre-processing registering the fNIRS channels over the native space of the participant will be performed (Bulgarelli et al., 2018).

Oxysoft software will be used to calculate, store and visualise real-time optical density haemodynamic response and 3D neuroimaging data, delineating the baseline and experimental periods (Richter et al., 2018). We will meet the hybrid reality training group again in 10 days (the number of days needed follows Martin et al. (2018))'s episodic memory research). They will have to identify similar safety risks to test their episodic memory of knowledge shared with fNIRS. In addition, behavioural data, including safety risk recognition accuracy and reaction time, will be collected throughout the experiment (Schaeffer et al., 2014).

Following Lawrence et al. (2018) approach, we will use the scalp coupling index to identify and exclude channels with poor optode-scalp contact. We will also exclude channels with SCI motion artefacts using the HOMER2 hmr Motion Correct Wavelet function, which implements a simplified form of the wavelet filtering algorithm. The temporal pre-processing will remove physiological noise of the haemodynamic response data, such as low-frequency drift and cardiac oscillations (band-stop filter: 0.06–0.16–0.8–1.8 Hz; 5th order Butterworth high-pass filter: 0.008 Hz). Artefacts will be detected as changes in the moving standard deviation greater than 3 μM using a 1-s sliding window (Bulgarelli et al., 2018), spline interpolation will be used to reduce motion artefacts.

The optical density will be converted to estimate changes in deoxygenated haemoglobin and oxyhaemoglobin via the application of the modified Beer-Lambert Law. Data will be exported and analysed by a general linear model (GLM) in the HomER2 package. GLM assumes a Gaussian error structure and a canonical haemodynamic response function. The onset and duration variables estimate the standardised beta coefficients within the control and non-control trials for each condition. The sign and magnitude of each beta coefficient will indicate the change in the haemodynamic response under each condition. The experiment's outcome and the control group will be assessed by analysis of variance (Baker et al., 2018). A multifactorial Plackett–Burman design experiment will also be used to study the impact of various factors (Table 1) that affect holographic training, e.g. the changes observed in haemodynamic response, time to locate the safety risks (Table 2 and Fig. 1).

Four Holographic hybrid reality training scenarios will be built, with one a general construction site and three specially designed for a building refurbishment, new building, and civil engineering scenario (risks related to the green features will also be shown).

One safety risk will be shown together with the spatial sound, the second half of the training will include the special smell, some risks from the real world will be holographically virtually labelled by AI object detection (Table 3).

4 Conclusion

The construction industry is a notoriously high-risk industry in many countries. On-site accidents have many negative consequences: absenteeism, project delays, permanent disability, and hugely personal, social, and financial costs. The proposed study will use an object detection approach to identify construction site safety risks to train a computer. In recent years, AI research on vision has been widely applied; for example, some cities in China use facial-recognition software to catch thieves at concerts. The construction industry is catching up, but the application of AI is limited by its ability only to recognise the existence of a human being or detect one type of protective equipment. The naked eye currently identifies most safety risks, but an element of human error is involved, and individuals have different standards for what constitutes a safety risk. Furthermore, the ageing labour force and escalating labour costs in the construction industry mean that novices may lack mentors for on-site

Table 1 List of hypotheses (AI object detection (**a**), Hololens (**b**), fNIRS and oxysoft (**c**), questionnaires (**d**), interview questions (**e**), risk detection record (**f**)

Hypothesis	Hypothesis details (theoretical background)	Method
The holographic hybrid reality for hazard detection on site		
H1	The holographic hybrid reality with AI object detection helps construction practitioners to identify safety risks on sites and provide a standardised benchmark for whether a hazard exists, i.e. the probability level of each hazard	a, b, e
Holographic hybrid reality knowledge sharing		
H2	Holographic hybrid reality can be an effective means for safety knowledge retention due to its high engagement in nature (Chalhoub & Ayer, 2018)	a, b, c, e, f
H2a	There is a decrease in the level of increase in oxyhaemoglobin and the level of decline in deoxygenated haemoglobin after holographic training as compared to baseline, indicating that trainees are more competent in safety risks identification (Pinti et al., 2019, forthcoming)	a, b, c
H2b	After holographic hybrid reality safety training, hazard identification will be easier, thus more activity evidenced in the left dorsolateral prefrontal cortex by neuroimaging (Heekeren et al., 2004)	a, b, c
Safety risks and brain activation		
H3	The areas activated by perceived high, medium and low construction hazards are similar to those activated by driving risks (Megías et al. (2018)	a, b, c, e
Safety knowledge sharing, retention and memory		
H4	Episodic memory of holographic safety training is strongly associated with episodic prospection and experience (Prabhakar & Hudson, 2019)	a, b, d, e
H5	Holographic virtual labelling and more than one sense engagement (spatial 3D sound and smell) enhance knowledge retention. Thus, the search for related safety risks are eased, as per the haemodynamic response results (Kelly & Heit, 2017), and less time is required	a, b, e, f
H6	A disconnection between memory and envisioning the similar exists, i.e. people remember holographic training but cannot identify the safety risks (Prabhakar & Hudson, 2019) (note 1)	a, b, e, f
H7	Holographic hybrid reality training triggers flashbulb memory, thus better knowledge retention (Kryklywy et al., 2013)	a, b, e, f
H8	Limited level of fear-related stimuli enhances the episodic memory of the safety knowledge shared (Li 2018a)	a, b, e, f
Physiological and cognitive factors that affect episodic memory of the knowledge shared		
H9	Deficits in episodic memory may contribute to unhealthy lifestyle such as uncontrolled eating (Martin et al., 2018), alcohol consumption, and poor sleep	a, b, d, e
H10	Visual intelligence is closely related to IQ (Sellers et al., 2015), thus workers with a higher IQ detect risks faster, lower mental load and have a higher accuracy rate	a, b, c, d, f

Note The phenomenon of disconnection should be minimal to ensure effective knowledge sharing

Table 2 Experimental schedule

Days	Experience with HoloLens training	No experience with HoloLens training
1–10	Pilot studies	
21–30	Group 1a: 4 building refurbishment practitioners	Group 1b: 4 building refurbishment practitioners
31–40	Group 2a: 4 road construction practitioners	Group 2b: 4 road construction practitioners
41–50	Group 3a: 4 new building practitioners	Group 3b: 4 new building practitioners
51–60	Group 4a: 4 building refurbishment practitioners	Group 4b: 4 building refurbishment practitioners
61–70	Group 1a: 4 road construction practitioners	Group 1a: 4 road construction practitioners
71–80	Group 1a: 4 new building practitioners	Group 1a: 4 new building practitioners

risk identification. AI object detection can help industry practitioners to provide an object risk analysis and planning.

AI hazard detection can help workplace safety officers identify on-site hazards and reduce the risk of human error. Given the considerable compensation costs for construction workers involved in construction accidents and the high salaries of construction workers, the proposed study can potentially make an enormous impact on the construction industry. The wearable computer HoloLens allows the visualisation of hazard labels identified on-site. This lightweight, convenient tool lessens the need for electrical wiring on-site, reduces reliance upon experienced mentors for hazard identification and allows novices to identify hazards or even ask for help from safety officers in the office, which overcomes the issue of an insufficient number of mentors on-site.

4.1 Academic Significance

Academically speaking, most current AI-based object detection applications involve car identification and criminal facial recognition. Indeed, many AI vision researchers around the world have been developing libraries of photos for training; for example, Professor Fei-Fei Li from Standford developed the Caltech 101 photo library with 101 categories of 40 to 800 images that can be downloaded from Caltech's website (http://www.vision.caltech.edu/Image_Datasets/Caltech101/#Description), and Google Open Images has benefitted researchers worldwide. Although there are many construction site photos online, there is no systematic photo database for construction site safety risks. The construction of a photo library can thus be beneficial to computer vision scientists and researchers studying construction safety risks.

Stage 1 (1-13 months): Construction site safety risk photo library construction and object detection training

1. An expert panel will be formed to advise the researchers.
2. Creative Commons photos will be collected and other photos will be taken from videos. We will invite construction practitioners and safety trainers via social media groups and emails to send us photos of construction risks.
3. Safety officers, trainers and experienced construction practitioners will be invited to identify relevant construction site safety risks in the photos through the Captcha developed by our research team, photos will then be categorised into different types of safety risks.

↓

Stage 2 (14-19 months): To build a holographic hybrid reality for safety training and safety risk identification.

Holographic AI hybrid reality testing on construction sites

1. The trained figures will be deployed to the HoloLens in our laboratory.
2. Safety experts, including safety managers and trainers, will be invited to use the HoloLens on new building construction, building refurbishment and road construction sites to validate the usefulness of detecting hazards on-site and provide comments regarding object detection via HoloLens.

Holographic AI hybrid reality used for safety training

One general construction site safety holographic hybrid reality and three specific construction site safety holographic hybrid reality training modules (1 for refurbishment, 1 for road construction and 1 for new building) will be made:

1. The trained AI photos will be deployed from Azure to HoloLens.
2. Holograms will be made by Holokit and some virtual safety risks spatial sound, etc. will be included.

↓

Stage 3 (20-24 months): Haemodynamic response and neuroimaging studies

48 new building construction, building refurbishment and road construction practitioners will be involved in the research. Safety training sessions will last no longer than 20 minutes for the highest level of attention (Li, 2018).

HoloLens group

The HoloLens group will receive HoloLens safety training on the first day. We will give them a quick tutorial on how to interact in the hybrid-reality environment, such as by gazing and tapping. They will be trained on one general construction site and one specific site relevant to their work (road/new building/building refurbishment). They will return 10 days later and will be asked to detect some similar safety risks.

On both days, they will wear the mobile fNIRS which measures the haemodynamic response and records activities in different parts of the brain, which can be visualised by neuroimaging.

The control group

The control group will receive no HoloLens safety training at all, they will wear the fNIRS when they are asked to identify safety risks on the only and the day we meet.

Analyses

Haemodynmaic response of different parts of the brain will be analysed by a general linear model, Analysis of variance and multi-factorial Design of Experiment (DOE). We will also study the neuroimaging with and in absence of safety risks among these two groups.

Fig. 1 Research methods

Table 3 Questions to be asked in the first meeting with control and HoloLens training groups (A, B, D), in the second meeting with the group undertaking holographic hybrid reality training (A6-7, B, C6, D), and HoloLens aids safety risks detection group (A1-4, C1)

(A) Participants' information to be collected before the experiment (Questionnaire)
A1. Age: ________________   A2. Years of experience: ___________on construction site_________ current type of job   A3. Occupation: A4. Types of work (circle the main type of work in the last 5 years)   New building (green building/non-green building) / Refurbishment of an existing building (with/without green features) / Road construction   A5. IQ score (WAIS-IV)   A6. Eating behaviour (Three-Factor Eating Questionnaire)   A7. Lifestyle (including alcohol, smoking, sleeping habit, social engagement) (Klaming et. al. 2017, p.531) over the ten days before we meet
(B) Instructions to the research participants (B1 will be asked before the experiment, B2 will be asked after the experiment)
B1. You are required to notify what safety risks you have seen once you identify them   B2. Could you please rank the level of risk for each of the risks that you observed?
(C) Interview questions asked for assessing the effectiveness of HoloLens
HoloLens aids safety risks detection group   C1. Do you think the HoloLens helped you detect the safety risks? And Question C3   *Safety training group*   C2. Do you think the HoloLens helped you in learning?   C3. Is there anything you think should be improved?   C4. Could you please describe your experiences in using HoloLens for safety training?   C5. Please rate the level of (i) surprise and (ii) fear that you perceived from the Holographic hybrid reality training (from 0 being the lowest to 100 being the highest)   C6. What risks did you see last time?
(D) Risk detection record (to be collected by the researchers in the first meeting with the control group and the second meeting for the HoloLens trained group)
D1. Sequence, safety risk recognition accuracy, and reaction time of the safety risks identified

In addition, although the virtual reality is an effective tool for safety training, the effectiveness of a holographic hybrid reality approach is unknown, and there is little research on the brain's response to construction site safety risks in terms of the parts of the brain involved. The proposed project will provide insight into which parts of the brain are involved in decision-making for risk identification, improve existing safety training, and even further the development of neuroscience-based safety training in the construction industry.

4.2 Practical Significance

Practically speaking, an element of human error exists in hazard identification, and what individuals perceive as risks may vary. The training results can provide a standard measure for what constitutes a hazard and can be used to implement a holographic hybrid reality for safety training. If workers learn from such training and incorporate it into their memory, it is expected that their haemodynamic response will differ from that of those who have not received training.

Acknowledgment This work described in this paper was fully supported by a grant from the Research Grants Council of the Hong Kong Special Administrative Region, China (Project No. UGC/IIDS15/E01/19).

References

Baker J.M., Bruno J.L., Gundran A., Hosseini S.M.H. & Reiss A.L. (2018) fNIRS measurement of cortical activation and functional connectivity during a visuospatial working memory task. *PLOS ONE, 13*, e0201486.

Bulgarelli C., Blasi A., Arridge S., Powell S., de Klerk CCJM, Southgate V., Brigadoi S., Penny W., Tak S. & Hamilton A. (2018) Dynamic causal modelling on infant fNIRS data: A validation study on a simultaneously recorded fNIRS-fMRI dataset. *NeuroImage, 175*, 413–424.

Burnett, P. (2008). Variable decision strategies, rational choice, and situation-related travel demand. *Environment and Planning a: Economy and Space, 40*, 2259–2281.

Chalhoub, J., & Ayer, S.K. (2018). Using mixed reality for electrical construction design communication. *Automation in Construction, 86*, 1–10.

Chan, A.P.C., Yang, Y., & Darko, A. (2018). Construction accidents in a large-scale public infrastructure project: Severity and prevention. *Journal of Construction Engineering and Management, 144*, 05018010.

Chen, J., Song, X., & Lin, Z. (2016). Revealing the "Invisible Gorilla" in construction: Estimating construction safety through mental workload assessment. *Automation in Construction, 63*, 173–183.

Curry, A. (2017). Neuroscience starts talking. *Nature, 551*, S81.

Curtin, A., & Ayaz, H. (2018). The age of neuroergonomics: Towards ubiquitous and continuous measurement of brain function with fNIRS. *Japanese Psychological Research, 60*, 374–386.

de Winkel, K.N., Nesti, A., Ayaz, H., & Bülthoff, H.H. (2017). Neural correlates of decision making on whole body yaw rotation: An fNIRS study. *Neuroscience Letters, 654*, 56–62.

Fang, Q., Li, H., Luo, X., Ding, L., Luo, H., Rose, T.M., & An, W. (2018). Detecting non-hardhat-use by a deep learning method from far-field surveillance videos. *Automation in Construction, 85*, 1–9.

Fishburn, F.A., Hlutkowsky, C.O., Bemis, L.M., Huppert, T.J., Wakschlag, L.S., & Perlman, S.B. (2019). Irritability uniquely predicts prefrontal cortex activation during preschool inhibitory control among all temperament domains: A LASSO approach. *NeuroImage, 184*, 68–77.

Golizadeh, H., Hon, C.K.H., Drogemuller, R., & Reza, H. M. (2018). Digital engineering potential in addressing causes of construction accidents. *Automation in Construction, 95*, 284–295.

Heekeren, H. R., Marrett, S., Bandettini, P. A., & Ungerleider, L. G. (2004). A general mechanism for perceptual decision-making in the human brain. *Nature, 431*, 859.

Incekara, F., Smits, M., Dirven, C., & Vincent, A. (2018). Clinical feasibility of a wearable mixed-reality device in neurosurgery. *World Neurosurgery, 118*, e422–e427.

Jahani, S., Fantana, A. L., Harper, D., Ellison, J. M., Boas, D. A., Forester, B. P., & Yücel, M. A. (2017). fNIRS can robustly measure brain activity during memory encoding and retrieval in healthy subjects. *Scientific Reports, 7*, 9533.

Jebelli, H., Hwang, S., & Lee, S. (2018). EEG signal-processing framework to obtain high-quality brain waves from an off-the-shelf wearable EEG device. *Journal of Computing in Civil Engineering, 32*, 04017070.

Kelly, L. J., & Heit, E. (2017). Recognition memory for hue: Prototypical bias and the role of labeling. *Journal of Experimental Psychology: Learning, Memory, and Cognition, 43*, 955–971.

Koulinas, G. K., Marhavilas, P. K., Demesouka, O. E., Vavatsikos, A. P., & Koulouriotis, D. E. (2019). Risk analysis and assessment in the worksites using the fuzzy-analytical hierarchy process and a quantitative technique—A case study for the Greek construction sector. *Safety Science, 112*, 96–104.

Kryklywy, J. H., Nantes, S. G., & Mitchell, D. G. V. (2013). The amygdala encodes level of perceived fear but not emotional ambiguity in visual scenes. *Behavioural Brain Research, 252*, 396–404.

Kwon, S., & Cha, S. (2017). CAPTCHA-based image annotation. *Information Processing Letters, 128*, 27–31.

Lai, L. W. C., Chau, K. W., & Lorne, F. T. (2019). "Forgetting by not doing": An institutional memory inquiry of forward planning for land production by reclamation. *Land Use Policy, 82*, 796–806.

Larson, C. (2018). China's AI imperative. *Science, 359*, 628–630.

Lawrence, R. J., Wiggins, I. M., Anderson, C. A., Davies-Thompson, J., & Hartley, D. E. H. (2018). Cortical correlates of speech intelligibility measured using functional near-infrared spectroscopy (fNIRS). *Hearing Research, 370*, 53–64.

Li, R. Y. M. (2018a). *An economic analysis on automated construction safety: Internet of things, artificial intelligence and 3D printing*. Springer.

Li, R. Y. M. (2018b) Multiple object detection with artificial intelligence for enhancing construction safety [keynote]. In: *2018 International Symposium on Big Data and Artificial Intelligence, 29–30 December 2018*, Hong Kong.

Li, R. Y. M. (2019). *Construction Safety Informatics*. Springer.

Li, R. Y. M., & Leung, T. H. (2019). Computer vision and hybrid reality for construction safety risks: a pilot study. In *Advances in Intelligent Systems and Computing Book Series (AISC)* (Vol. 1027, pp. 17–22).

Li, R. Y. M., Chau, K. W., & Ho, D. C. W. (2017). Dynamic panel analysis on construction accidents in Hong Kong. *Asian Journal of Law and Economics, 8*, 1–14.

Liao, C.-W., & Chiang, T.-L. (2016). Reducing occupational injuries attributed to inattentional blindness in the construction industry. *Safety Science, 89*, 129–137.

Lingard, H., Blismas, N., Harley, J., Stranieri, A., Zhang, R. P., & Pirzadeh, P. (2018). Making the invisible visible: Stimulating work health and safety-relevant thinking through the use of infographics in construction design. *Engineering, Construction and Architectural Management, 25*, 39–61.

Loos E., Egli T., Coynel D., Fastenrath, M., Freytag, V., Papassotiropoulos, A., de Quervain, D.J.F., & Milnik, A. (2019) Predicting emotional arousal and emotional memory performance from an identical brain network. *NeuroImage*.

Loosemore, M., & Malouf, N. (2019). Safety training and positive safety attitude formation in the Australian construction industry. *Safety Science, 113*, 233–243.

Luo, X., Li, H., Cao, D., Dai, F., Seo, J., & Lee, S. (2018). Recognising diverse construction activities in site images via relevance networks of construction-related objects detected by convolutional neural networks. *Journal of Computing in Civil Engineering, 32*, 04018012.

Martin, A. A., Davidson, T. L., & McCrory, M. A. (2018). Deficits in episodic memory are related to uncontrolled eating in a sample of healthy adults. *Appetite, 124*, 33–42.

Matthews, D. (2018). Science goes virtual. *Nature, 557*, 127–128.

Megías, A., Cándido, A., Maldonado, A., & Catena, A. (2018). Neural correlates of risk perception as a function of risk level: An approach to the study of risk through a daily life task. *Neuropsychologia, 119*, 464–473.

Milton, J., Solodkin, A., Hluštík, P., & Small, S. L. (2007). The mind of expert motor performance is cool and focused. *NeuroImage, 35*, 804–813.

Minitab. (2019). Designing an experiment. https://support.minitab.com/zh-cn/minitab/18/getting-started/designing-an-experiment/

Namian, M., Albert, A., Zuluaga, C. M., & Jaselskis, E. J. (2016). Improving hazard-recognition performance and safety training outcomes: Integrating strategies for training transfer. *Journal of Construction Engineering and Management, 142*, 04016048.

Pandit, B., Albert, A., Patil, Y., & Al-, A. J. (2019). Impact of safety climate on hazard recognition and safety risk perception. *Safety Science, 113*, 44–53.

Pinti, P., Tachtsidis, I., Hamilton, A., Hirsch, J., Aichelburg, C., Gilbert, S., & Burgess, P. W. (2019 forthcoming). *The present and future use of functional near-infrared spectroscopy (fNIRS) for cognitive neuroscience*. Annals of the New York Academy of Sciences.

Prabhakar, J., & Hudson, J. A. (2019). Past is prologue: The role of memory retrieval in young children's episodic prospection. *Journal of Experimental Child Psychology, 177*, 17–35.

Raptis, G. E., Fidas, C., & Avouris, N. (2018). Effects of mixed-reality on players' behaviour and immersion in a cultural tourism game: A cognitive processing perspective. *International Journal of Human-Computer Studies, 114*, 69–79.

Ren, H., Wang, M.-Y., He, Y., Du, Z., Zhang, J., Zhang, J., Li, D. & Yuan, Z. (2018) A novel phase analysis method for examining fNIRS neuroimaging data associated with Chinese/English sight translation. *Behavioural Brain Research.*

Richter, H.O., Forsman, M., Elcadi, G.H., Brautaset, R., Marsh, J.E. & Zetterberg, C. (2018) Prefrontal cortex oxygenation evoked by convergence load under conflicting stimulus-to-accommodation and stimulus-to-vergence eye-movements measured by NIRS. *Frontiers in Human Neuroscience, 12.*

Schaeffer, J. D., Yennu, A. S., Gandy, K. C., Tian, F., Liu, H., & Park, H. (2014). An fNIRS investigation of associative recognition in the prefrontal cortex with a rapid event-related design. *Journal of Neuroscience Methods, 235*, 308–315.

Sellers, K. K., Mellin, J. M., Lustenberger, C. M., Boyle, M. R., Lee, W. H., Peterchev, A. V., & Fröhlich, F. (2015). Transcranial direct current stimulation (tDCS) of frontal cortex decreases performance on the WAIS-IV intelligence test. *Behavioural Brain Research, 290*, 32–44.

Son, H., Choi, H., Seong, H., & Kim, C. (2019). Detection of construction workers under varying poses and changing background in image sequences via very deep residual networks. *Automation in Construction, 99*, 27–38.

Tang, R., Fong, S., Deb, S., Vasilakos, A. V., & Millham, R. C. (2018). Dynamic group optimisation algorithm for training feed-forward neural networks. *Neurocomputing, 314*, 1–19.

Wang, D., Tang, J., Zhu, W., Li, H., Xin, J., & He, D. (2018). Dairy goat detection based on Faster R-CNN from surveillance video. *Computers and Electronics in Agriculture, 154*, 443–449.

Winge, S., Albrechtsen, E., & Mostue, B. A. (2019). Causal factors and connections in construction accidents. *Safety Science, 112*, 130–141.

Zhong, Y., Han, X., & Zhang, L. (2018). Multi-class geospatial object detection based on a position-sensitive balancing framework for high spatial resolution remote sensing imagery. *ISPRS Journal of Photogrammetry and Remote Sensing, 138*, 281–294.

Chapter 9
Predicting Housing Prices in Hong Kong Based on AI Interpreted Sentiment in Social Media, Health and Sustainability Factors: A White-box AutoML Research Agenda

Rita Yi Man Li and Kwong Wing Chau

Abstract In recent years, automation via deep learning/machine learning has flourished. As such, automated valuation will likely become a major task in valuation over the coming years. While most studies in industry/academia utilise multivariate regression for housing price predictions, we aim to introduce explainable AutoML models (H_2O AutoML and Google Cloud Platform (GCP AutoML Tables) for housing price predictions, which will be the first of its kind. The objectives are as three folds: 1. predict housing prices based on the following factors: (i) sentiment, as reflected in (ia) animal spirits (market sentiment) interpreted and constructed using social listening tools with artificial intelligence, and (ib) a sentiment index constructed based on Google searches; (ii) health-related factors, including COVID-19 and radio base stations; (iii) sustainability factors, i.e., the urban heat island effect and polychlorinated dibenzo-*p*-dioxins (PCDD); (iv) the number of years left on a land lease; and (v) the impact of precarious social events on housing prices. 2. Study the impact of these factors and compare their relative importance on housing price predictions. 3. Predict housing prices via (i) H_2O Auto ML and (ii) GCP AutoML Tables and compare their results. The first stage includes a systematic literature review and content analysis of big data housing price research for the past ten years in Science Direct, Emerald, Sage and Taylor, and Francis, to uncover the current state-of-the-art research in housing prices and revise/update the variables to be included in our prediction model. Fifty-five highest housing price transaction records in Hong Kong will be used for the housing price predictions. Housing estate name, price, date of transaction, gross floor area, building completion date, and floor level will be downloaded from the Economic Property Research Center (EPRC). We will then reformat the data so that it can be fitted to AutoML. Second, we will obtain the sentiment of 55 housing estates and the "Hong Kong's housing market", as reflected in social media, by using

R. Y. M. Li (✉)
Sustainable Real Estate Research Center/Department of Economics and Finance, Hong Kong Shue Yan University, North Point, Hong Kong

K. W. Chau
Department of Real Estate and Construction, The University of Hong Kong, Pok Fu Lam, Hong Kong

R. T. M. Li et al. (eds.), *Current State of Art in Artificial Intelligence and Ubiquitous Cities*, https://doi.org/10.1007/978-981-19-0737-1_9

keywords to search on a Natural Language Processing platform. Third, PCDD will be obtained from the Environmental Protection Department, while radio base station locations will be obtained from the Office of the Communications Authority. The number of years left on the housing estates' land lease will be obtained from the Land Registry Department, with dummy variables being assigned in the prediction models to represent the precarious protest period and the number of COVID-19 cases obtained from Statista to study their impact on housing prices. Moreover, the remote sensing data of housing estates obtained from Landsat 8 (an American Earth observation satellite) can be downloaded through the United States Geological Survey (USGS). The land surface temperature, computed by ENVI 5.3 software, will be compared with nearby areas of vegetation to compute the extra heat caused by the "urban heat island effect". Big data collection will allow us to utilise H_2O AutoML and GCP Cloud AutoML for housing price predictions. We will apply H_2O AutoML to select the best machine learning algorithms. Unlike traditional machine learning algorithms, AutoML minimises parameter adjustments and automatically tunes all parameters according to embedded algorithms. The results will then be compared with the results from GCP Cloud AutoML. We will also select the model with the highest efficacy, which will reveal the relative importance of factors and show the coefficients of the factors on housing price predictions.

Keywords Housing price · Prediction · AutoML · Sentiment · Sustainability · Health · Automated valuation

1 Introduction

Housing price research is one hot topic in real estate market for many years. An econometric model constructed by Yang et al. (2019) found that, compared to low-end property buyers, high-end property buyers have a predilection for premium amenities such as natural beauty. By using 2,175,911 housing transactions data, including housing transactions in Tseung Kwan O district in Hong Kong, our results show those housing estates near landfills may not lower housing prices if landfill wind does not sweep across these units (Li & Li 2018). Zheng et al. (2015) found that market liquidity was significantly priced in cross-sectional asset pricing models.

The banking industry has utilised automated valuation for the preliminary estimations of housing prices. Many valuations are done using multivariate regression models or trend analysis. We attempt to utilise a novel approach (i.e., AutoML) to make these predictions. Second, previous housing price research studied the impact of socioeconomic factors such as unemployment rate, population, real interest rate, distance to the subway station, housing supply, and money supply (Li & Li 2018). In recent years, automation via artificial intelligence (AI) has flourished. As such, many tasks are now much easier to perform. At present, while most automated valuations utilise regression analysis to facilitate predictions, we aim to introduce H_2O AutoML and GCP AutoML—two types of AI models used for predictions.

To achieve this, apart from including certain factors that are known to impact housing prices (e.g., size), we will also include four groups of factors that are lacking from existing research: (i) sentiment, as reflected in big data social media and when searching for terms such as "Hong Kong housing market" and the names of individual housing estates; (ii) sustainability/health-related factors, including COVID-19, the urban heat island effect, polychlorinated dibenzo-*p*-dioxins (PCDD), and radio base stations; (iii) number of years left on a land lease; and (iv) the impact of precarious social events on housing prices. We will study these factors impacts and compare their relative importance on housing price predictions. The results will benefit valuers, surveyors, mortgage providers, existing homeowners, potential buyers and investors in knowing the impact of these factors on housing prices and utilise the algorithm for valuation and housing price prediction in the long run.

2 Literature Review

2.1 Different Types of Sentiment Proxies Impact Housing Price

As early as 1936, economist John Maynard Keynes suggested that "animal spirits" affect consumer confidence. Today, "animal spirits" refers to the market sentiment that affects investors' acts or attitudes within a market (Chevapatrakul et al., 2019; Ruan et al., 2020) and "animal spirits" have been considered to impact asset valuations. In the housing market, by using *housing transaction volume and amount* as a market sentiment proxy, Hui et al. (2017) found that buyer–seller sentiment negatively affected housing returns, whereas the developer sentiment's effect was positive.

Hui and Wang (2014) found that market sentiment motivated housing demands based on a sentiment index built using *transaction records*. Zhou (2018) found that housing market returns rise with an increase in coexistent sentiment. Here, the sentiment proxy included newly released residential land supply, the housing sector investment divided by total real estate, the log of the median holding period of house sellers, the total area of transacted houses divided by the total area of houses available for sale, and the return of the small-house index minus the return of the big-house index.

The advent and popularity of social media have transformed the web from publisher-created content to user-created content (Ghani et al., 2019). Social media provides a platform for people to share their thoughts on various issues (Sailunaz & Alhajj, 2019), such as construction and occupational safety (Song et al., 2022; Yao et al., 2021), AI used for the construction and urban planning activities (Regona et al., 2022; Yigitcanlar et al., 2020) and is a big data hub (Ghani et al., 2019). For example, around one-third of the world's population use social media; Facebook alone had 2.23 billion monthly users as of 2018. Thus, information shared on these platforms affects users' perceptions and decision-making processes—as per social influence theories (Wiese et al., 2020). Nevertheless, carefully fusing vast amounts of opinions is challenging. Social listening extracts data (e.g., conversations, comments, etc.) and performs sentiment analysis via natural language processing, giving rise to new business in the market (Ducange et al., 2019). Ruan et al. (2020) suggest that AI estimates investor sentiment value directly without a pre-set emotional dictionary, thus reducing subjective bias.

Apart from social listening, Google has become a tool for predicting economic and financial markets. Updyke et al. (2017) stated that using Google econometrics, i.e., Google Trends, which analyses the popularity of top search queries in Google Search across various regions and languages, is used to predict economic variables (González-Fernández & González-Velasco, 2018). Khan et al. (2019) constructed a household stock market investors' sentiment index using Google Trends data. They confirmed that the Google Trends sentiment index had causal effects and strongly correlated with the stock market index.

Niesert et al. (2019) found that Google Search data is most helpful when the series under investigation is directly related to an individual's situation and is closely linked with a specific search. For example, many unemployed people know in advance that they may become unemployed; this knowledge becomes a predictable online search behaviour. By studying Google Trends data from 2004–2016, Li et al. (2016) found that the keyword "home automation" was recorded as the most extensive search entry in Bengaluru (India), reflecting a keen interest in home automation in this location.

However, despite previous research utilising different sentiment proxies to study sentiment's impact on housing prices, none have utilised AI interpreted sentiment in housing price predictions. Likewise, no research has used Google Trends as a sentiment proxy for housing price predictions via AutoML. Thus, we hypothesise:

H1a: Higher positive (negative) animal spirits (sentiment) expressed in social media (interpreted by AI via a social listening tool) positively (negatively) impacts housing prices.

H1b: More Google searches for "housing prices in Hong Kong" positively correlate to housing price predictions.

2.2 Health and Housing Prices

Previous research on the weekly transaction prices of 44 housing estates during the severe acute respiratory syndrome (SARS) epidemic showed that irrationality was associated with a significant drop in housing prices (Wong, 2008). On 22 January 2020, two COVID-19 cases in Hong Kong were reported, opening Pandora's Box. COVID-19 began as a health crisis, and however, it quickly became an economic one (McKee & Stuckler, 2020). We speculate that the number of COVID-19 cases negatively impacts housing prices.

Radiofrequency radiation (RFR), used for wireless and intelligent building technologies, is rapidly increasing and has been known to cause cancer and early childhood developmental problems. The World Health Organization (WHO) classified

100 MHz to 300 GHz RFR as a possible human carcinogen and cause of glioma (Clegg et al., 2019). In Hong Kong, with an increase in mobile communication networks, there has been an increase in demand for high-speed mobile broadband and data services. Radio base stations have been installed on building rooftops to improve mobile network coverage, data usage, and speed. More people have recently reported illnesses (including cancer) due to radiation being released by radio base stations, which has caused increased concern. Although there has been a rapid increase in RFR exposure, as well as scientific evidence about its harmful effects, the terms "building science" and "RFR" are rarely used in the same sentence (Clegg et al., 2019).

That said, one exception is a recent case reported in Stockholm, where high RFR levels in an apartment were found, with two base stations nearby (Hardell et al., 2018). We conjecture that people have health concerns regarding RFR, which negatively impacts housing prices. Thus, we hypothesise:

H2: COVID-19 (H2a) and radio base stations (H2b) negatively impact housing prices.

2.3 Sustainability and Housing Prices

Densely populated urban areas record a higher ambient temperature than surrounding rural and suburban areas. This is known as the urban heat island effect. Urban overheating is caused by anthropogenic heat, canyon radiative geometry, reduction of evaporative surfaces, and reduced turbulent transfer in urban environments. It is well documented that more than 400 cities worldwide suffer from the effect, which can raise average temperatures by over 4–5 °C, peaking at over 10 °C. Increased ambient temperatures elevate cooling energy consumption, heat-related mortality, morbidity, increased pollutant concentration and causes discomfort with urban quality (Mathew et al., 2019). As such, we speculate that the urban heat island effect negatively impacts property prices. Nevertheless, no research has looked at this to the best of our knowledge, except for the authors' pilot research focusing on one housing estate in Hong Kong.

Similarly, the impact of externalities due to different air pollutants on property prices is well documented: CO, NO_2, PM2.5, and PM10 significantly negatively correlate with housing prices, but there is an insignificant relationship between housing prices SO_2 and O_3 (Mei et al., 2020). While previous research has found that polychlorinated dibenzo-*p*-dioxins (PCDD), a type of air pollutant, may lead to children suffering from global DNA hypomethylation (Xu et al., 2019), research on the relationship between PCDD and housing prices is scarce (if not absent). Therefore, we hypothesise:

H3: The urban heat island effect (H3a) and PCDD (H3b) negatively impact housing prices.

2.4 The Impact of Precarious Social Events on Housing Prices

In theory, if a riot's effect on perceived amenities is small and short-lived, any impact would be transitory. Furthermore, with large transfers of federal or state resources, a riot's effect might be negligible (Collins & Margo, 2007). However, if an intense conflict area causes a sustained decline in perceived amenities in one location relative to others, we should be able to detect a decline in property values, a reduction in investment and emigration. Besley and Mueller (2012) found that six months after the conflict in Northern Ireland, the event led to a more significant negative impact on housing prices. From June 2019 to 2020, Hong Kong became overwhelmed by months of social events. We hypothesise that:

H4: Homeowners slash prices after social events, as reflected in housing price predictions.

2.5 The Impact of the Number of years Left on a Land Lease on Housing Prices

A land lease is a contract between a landlord and tenant for a specific period. Within this contract, the tenant holds the right to possession and use. When the lease expires, possession rights and other interests are returned to the landlord or their successor. Fewer years left on a land lease implies earlier risks/earlier extra land premiums or even possession rights being returned to the landlord. Further, Gautier and Vuuren (2019) found that privately owned land is 11%–13% more expensive than observably equivalent houses with a non-prepaid land-lease contract.

On 9 January 2020, the Hong Kong Economic Journal reported that many land leases in Hong Kong will expire in 2047. In response, WaFang suggested that we avoid purchasing housing estates with leases expiring in 2047 or earlier to reduce the risk of extra land premiums (see http://tiny.cc/f9g8iz). As such, we speculate that housing with fewer years left on a land lease may be less attractive to home buyers, resulting in lower housing prices. However, no research on land-lease terms and housing prices exists in Hong Kong, not its impact on predicting housing prices. Thus, we hypothesise:

H5: The fewer years left on a land lease, the lower the housing price.

3 Research Method

3.1 Housing Price Prediction and Auto-Machine Learning (AutoML)

A feedforward neural network model was adopted in Li et al. (2019), consisting of the input, middle hidden, and last output layers, with the product of input elements (ai) and weights (wij) fed to summing junctions with a bias (bj) of neurons as X = $(\sum_{i=1}^{n} w_{ij})$ + bj where the sum passes through the transfer function F to generate output: F(X) = uj = F[$(\sum_{i=1}^{n} w_{ij}\ a_i)$ + bj]. An activation function estimates the feature importance of factors that affect construction safety knowledge shared on various tools. In Li et al. (2018), by using *walled buildings* as a proxy for towers experiencing the urban heat island effect, our artificial neural network model showed that the urban heat island effect lowered housing prices. Li *et al.* (2016a) utilised SANN AutoML to study the impact of moral hazards on accident compensation. We only used H_2O AutoML and GCP AutoML Tables for predicting housing prices in our pilot studies.

For both novices and experts in machine learning, model selection and hyper-parameter optimisation is a trial-and-error procedure. Thus, an automated objective approach such as AutoML is needed (Zhang et al., 2020). Based on transaction count, size, market capitalisation, and the Google Trends search volume index, etc., Chen et al. (2020) used AutoML to predict Bitcoin prices. Johann et al. (2016) utilised the Statistica Automated Neural Network (SANN) to select the best AutoML models for soil moisture prediction and found that RBF was less dependent on the training–testing data. Nevertheless, H2o.ai AutoML and GCP AutoML were not used for housing price predictions.

3.2 H_2O Explainable AutoML for Prediction

We expect to conduct data analysis and handle missing data through the EM algorithm. The next step will be to examine data stationarity. An Augmented Dickey-Fuller test will be performed using $\Delta Y_t = (p-1)Y_{t-1} + u_t,\ p-1 = \delta$. Then, we will implement H_2O AutoML and GCP AutoML, the two most popular AutoML approaches (research.aimultiple.com/automl-comparison), to fulfil our three objectives.

H_2O AutoML allows developers with limited machine learning knowledge to develop machine learning models (Lee et al., 2019). H2o.ai is an explainable AutoML algorithm, i.e., a white-box model. Sensitivity allows us to see what happens to predicted housing values when a specific value of a factor changes. The Shapley value will allow us to see the impact of factors regarding the direction and value (H2o.ai, 2021) to be fulfilled in objective 2. H_2O requires between 1,000 and 100,000,000 rows of data and at least two but no more than 1,000 columns. Unlike traditional machine learning, AutoML minimises the number of parameter adjustments and automatically tunes all parameters as per the embedded algorithms in order to select the best model based on the machine learning algorithms: XGBoost, distributed random forest, generalised linear model, gradient boosting machine, and stacked ensemble (Jang, 2019).

3.2.1 Extreme Gradient Boosting (XGBoost)

Parallelises gradient boosted trees' growth to reduce time, speed up optimisation procedures, and grow trees. Each tree is approximated by numerous regression functions fi(x) (Chen et al., 2020). Feature importance is evaluated according to the average gain during node splitting (Zhai & Chen, 2018). XGBoost's objective function is computed as $\text{Obj}(\theta) = \sum_{k=1}^{k} \vartheta(f_k) + \sum_{i=1}^{n} L(y_i, \hat{y}_i)$ where the first part shows the sum of each tree's complexity, and the second part illustrates the model's training loss. Tree complexity is computed as $\vartheta(f_k)=\gamma T + 0.5\aleph||w||^2$, where γ shows the complexity parameter, T is the leaf nodes number, $\aleph$ is the fixed coefficient, and $||w||^2$ is the l_2-norm of leaf weight. Accelerating the convergence training process and adding a regularised term to prevent overfitting, XGBoost has been used for stock price predictions (Jiang et al., 2020).

3.2.2 A Gradient Boosting Machine (GBM)

GBM is a homogeneous ensemble learner, in which trees grow sequentially, and later trees grow depending on the results of previous trees. GBM requires tuned hyperparameters and applies a grid search to achieve reasonable accuracy (Tama & Comuzzi, 2019). Ho et al. (2021) found that GBM outperformed support vector machines in prediction but did not compare the results of AutoML.

3.2.3 A Distributed Random Forest (DRF)

DRF generates a forest of regression trees for the data. Each tree is a weak learner built on row and column subsets. Regression takes an average across all trees to make a prediction. DRF maps factors in lexicographic order to a name lookup array with integer indices for a categorical response column (H2O.ai, 2021).

3.2.4 A Generalised Linear Model (GLM)

GLM allows for data other than normal distributions. It displays a curvilinear relationship between a response variable Y and explanatory variables x_i: $Y = a_0 + a_1x_1 + a_2x_2 + \ldots. + a_kx_k + \ldots + \varepsilon$, where a_0 is the regression coefficient for the intercept and a_k are the regression coefficients of explanatory variables 1 through k, computed using the training data and explanatory variables x_k, where x is a linear or higher power, and ε is the error term. GLM has been used to study the percentage of tree cover in remote sensing data (Schwarz & Zimmermann, 2005).

3.2.5 A Stacked Ensemble

A stacked ensemble combines the models above to achieve higher prediction accuracy and reduce generalisation error. It trains L base learners in the training set, performs k-fold cross-validation on base learners, and cross-validates the predicted values p1, …pL. The N cross-validated predicted values from each of the L algorithms is combined to form new N rows in the training set, which are then multiplied by the L feature to form a matrix z. The matrix, together with the original response vector y, forms level one data. By training the meta-learning algorithm on level one data (y = f(z)), the ensemble model consists of L base learning and a meta-learning model (Boehmke & Greenwell, 2020). Previous research has shown that a stacked ensemble outperforms multilayer perceptron in PM2.5 and stock price prediction (Jiang et al., 2020; Zhai & Chen, 2018).

3.2.6 GCP explainable AutoML Tables

GCP AutoML Tables automatically computes global feature importance for the trained model. A feature with a higher value means a more significant impact on a model's predictions. If a feature's contribution has a low value, it can be excluded from future training. The local feature importance, i.e., a score showing how much and in which direction each feature influenced the prediction for a single example, is also shown. The What-If Tool can be used to see how our model performs across a larger dataset and visualises values' attribution (Google, 2021).

AutoML Tables trains multiple model architectures simultaneously to quickly determine the best model without iteration over numerous model architectures. Model tests include GCP AutoML's AdaNet algorithm, Gradient Boosted Decision Tree, feedforward deep neural networks, and linear regression models.

3.2.7 AdaNet

AdaNet creates base learners to be included in the final ensemble. It learns network structure and its weights, and optimises the common loss function. At every iteration, the trained base learners evaluate their ability to minimise the AdaNet objective, with the best one being included: $F(w) = \sum_{i=1}^{m} \varnothing(\sum_{j=1}^{N} w_i h_i(x_i), y_i) + \sum_{j=1}^{N}(\gamma_r(h_j) + \rho|w_j|$ where wj is the weight of model j's contribution (Adanet, 2018).

3.2.8 The Gradient Boosted Decision Tree

The Gradient Boosted Decision Tree handles the multicollinearity between explanatory variables. It ranks the influential factors in terms of predicted value. The training set is randomly extracted from the original sample set $\{x(t), y(t)\}_{t=1}^{m}$ which consists of m samples. t training samples are extracted from the original sample set each time by bootstrapping, with selected training samples being placed back after each training (1/m). After training b times, b sample sets are selected. Regression coefficients φ_j of each variable will be calculated during training. A final function f(x) is obtained via averaging and adding (Wu et al., 2020).

3.2.9 Feedforward Deep Neural Networks and Stacked Ensemble

In feedforward deep neural networks, each non-output layer has a bias unit that serves as an activation threshold for neurons in the subsequent layer. Each neuron receives a weighted combination α of the *nl* outputs of neurons in the previous layer l as an input: $\sum_{i=1}^{n_l} w_i x_i + b$, where w_i and b are xi's weight and bias, respectively. The weighted combination α is transformed via an activation function f, and an output signal f(α) is relayed to neurons in the subsequent layer (Orimoloye et al., 2020). A stacked ensemble combines the aforementioned models for more accurate prediction.

GCP AutoML regression performs best with minimum rows of data at least 200 × the number of features, and less than 1,000,000 rows or 300 columns. A quality model (accuracy, low RMSE, etc.) can be achieved without manual feature engineering, model selection, and ensembling. The final model hyperparameters and objective values used during model validation can be seen in logging (Waring et al., 2020). By default, AutoML Tables uses 80% of data for training, 10% for validation, and 10% for testing. These can be manually edited if needed (Google, 2021).

3.3 *Efficacy and Accuracy of the Model Comparison*

The efficacy and accuracy of AutoML will be measured by (1) the root mean squared error (RMSE) = $\left(\sum_{i=1}^{r} (\mu_{exp} - \mu_{pred})^2 / r\right)^{0.5}$ for the average magnitude of the error, and (2) the mean absolute error (MAE) = $\sum_{i=1}^{r} \left|\mu_{exp} - \mu_{pred}\right| / \mu_{exp}$ where μ_{exp} is the experimental viscosity, and μ_{pred} is the predicted viscosity (Yadav et al., 2020). MAE measures the average absolute value of difference between the predicted and actual values. Smaller RMSE and MAE values indicate higher prediction accuracy.

3.4 *Remote Sensing for Measuring the Urban Heat Island Effect*

Obtaining the land surface temperature is a primary step in the analysis of the urban heat island effect. In other words, the relative warmth of urban surfaces compared to non-urbanised areas/vegetation (Mathew et al., 2019). Remote sensing techniques are often used to study variations in urban heat islands among different land covers, as they can accurately monitor the spatiotemporal changes of urban heat islands (Wang et al., 2019).

3.5 *Data*

3.5.1 Housing Price Data

We will collect data from all variables, as listed in Table 1. Unlike traditional econometric models, which cannot handle many variables, H_2O is designed to handle big data. In 2021, we will collect housing price data from the top 55 housing estates in Hong Kong between 2004–2022, with the highest transaction number as stated by Midland Realty—one of Hong Kong's most prominent property agencies. The text data file of transaction price and date, gross floor area, building completion date, and floor level will be downloaded from the EPRC. We will reformat the data from a text

Table 1 Summary of target and control variables

Target testing variables (hypothesis)	Literatures	Data source
Social Media sentiment (D)[1, 3], e.g. YouTube, Hong Kong Goden, Facebook, Instagram, Uwants etc. (H1a)	Hui et al. (2017), Hui and Wang (2014), Zhou (2018), Ducange et al. (2019), Ruan et al. (2020)	Social listening company (e.g. K-matrix) (H1a)
Google Search index (M)[1, 2, 3] (H1b)	González-Fernández and González-Velasco (2018), Khan et al. (2019), Updyke et al. (2017)	Google Trends (H1b)
COVID-19 (D)[3] (H2a)	Wong (2008), McKee and Stuckler (2020)	Statista (H2a)
Mobile base stations location (dummy variable)[1, 2, 3] (dummy variable will be assigned to the highest 5 units with based stations in the tower**) (H2b)	Clegg et al. (2019)	OCFA (H2b)
Heat island effect (S)[1, 2, 3] (H3a)	Mathew et al. (2019)	USGS (H3a)

(continued)

Table 1 (continued)

Target testing variables (hypothesis)	Literatures	Data source
PCDD (M) (control variables: SO_2, CO, O_3, NO_2, PM2.5, PM10 (D)[1, 2, 3] (H3b)	Mei et al. (2020), Xu et al. (2019)	Environmental and Protection Department (H4b)
Precarious social events (dummy variable after 2020)[3] (H4)	Besley and Mueller (2012), Collins and Margo (2007)	Local news (H5)
Number of years left on a land lease (Y)[1, 2, 3] (H5)	Gautier and Vuuren (2019)	Land Registry (H6)
Control variables		
Different age groups' population (S), marriages (M), female and male population (S), unemployment rate (Q) income (Q, real payroll indices), net migration (Y) [1, 2, 3]	Lin et al. (2014), Green and Lee (2016), Costello et al. (2019)	Census and Statistics Department/Moody's Analytics
Hang Seng Stock index (D)[1, 2, 3]	Li and Li (2018)	Yahoo! Finance
Interest rate (D)[1, 2, 3]	Li and Li (2018)	HKMA
Global money supply (M)[1, 3]	Li and Li (2018)	OECD Data
Exchange rate (Hong Kong to RMB) (D)[1, 2, 3]	Li and Chau (2016)	Investing.com
Housing price, holding period, transaction date (D), building age (Y), floor, net floor area/minimum travel time to CBD[1, 2, 3]	Lin et al. (2014)	EPRC (H1–H6)/Google Maps
Shenzhen (one of the closest cities to Hong Kong) housing price index (M) [1, 2, 3]	Yang et al. (2018)	Fang.com
Stamp Duty Policy (dummy variable will be assigned)[2]	Wu et al. (2017)	Inland Revenue Department

Notes for the Table
[1]Model 1 utilises data of the previous ten years due to social media data availability
[2]Model 2 examines the impact from 2004 to 2022 due to Google Trend data availability
[3]Model 3 studies the impact from 2018 to 2022 due to relative short data like Covid-19
Models 1–3 will include 1a) linear and 1b) semi log forms

file so that it can be fitted to H_2O AutoML. Next, we will remove duplicate units, missing data, and outliers (i.e., the top and bottom 0.1% of data) as per Leung and Tsang (2013). Transactions before building completion will be excluded to rule out presale cases, as these are associated with money paid back by developers, making it difficult to estimate the actual transaction price (Zheng et al., 2015). This also avoids transactions between relatives or specially designed units.

3.5.2 Land Surface Temperature Collection

The latitude and longitude of the housing estates (the central point and four other points along the diagonal) will be obtained from Google Maps to determine where the land surface temperature data needs to be retrieved. Landsat 8 provides space-based moderate-resolution remote sensing data, which can be freely downloaded from the USGS website (earthexplorer.usgs.gov). Band 10 of Landsat 8 is relatively stable and thus will be used to retrieve the land surface temperatures. Cloud-free images will be used to retrieve Radiative Transfer Equation-based land surface temperatures (Sekertekin, 2019).

After collecting satellite images and vector maps of Hong Kong, image pre-processing is required since Landsat 8 files only contain independent single-band images; the final satellite images will be combined using layer stacking software ENVI 5.3 to form multi-band images. Geometric correction will be used for radiation correction. According to the latitude and longitude of the housing estates, Hong Kong vector maps will be used to clip 55 housing estates using ENVI 5.3.

The radiative transfer equation (RTE) will compute land surface temperatures from 2004–2022. According to their highly cited paper, Yu et al. (2014) simplified the radiative transfer equation to express the apparent radiance received by a sensor: $B_i(T_i) = I_i^{\uparrow} + \tau_i\theta[\varepsilon_i B_i(T_s) + (1\text{-}\varepsilon_i)I_i^{\downarrow}]$. Where $B_i(T_i)$ represents the radiance received by sensor channel i with brightness temperature T_i; details regarding the TIRS' calibration is stated on the USGS website. $B_i(T_s)$ is the ground radiance, $\tau_i\theta$ records the atmospheric transmittance for channel i when the view zenith angle is θ, and TIRS is treated as the nadir viewing angle since the view angle is no more than 7.5 degrees. ε_i refers to the surface emissivity of channel i, $I_i^{\downarrow}$ and $I_i^{\uparrow}$ represents the downwelling and upwelling path radiance, respectively.

According to Plank's law, $B_i(T_s)$=2 h $c^2/\gamma_i^5 * (expexp\left(\frac{hc}{\gamma k T_s}\right) - 1))$, where T_s is the land surface temperature, c is the light speed (c = 2.9979 × 108 m/s), h represents the Planck constant (h = 6.6261 × 10-34Js), k is the Boltzmann constant (k = 1.3806 × 10–23 J/K) and γ_i is the effective band wavelength for band I, which is defined as $\gamma_i = \int_{\gamma_{1,i}}^{\gamma_{2,i}} f_i(\gamma)\gamma d\gamma / \int_{\gamma_{1,i}}^{\gamma_{2,i}} f_i(\gamma) d\gamma$ where $f_i(\gamma)$ is the spectral response function for the corresponding band, $\gamma_{1,i} and \gamma_{2,i}$ represents the lower and upper boundary of $f_i(\gamma)$ (Yu et al., 2014)

The surface temperature T_s = C1/γ_i In(C2/$\gamma_i^5(B_i(T_i) - I_i^{\uparrow} - \tau_i(1-\varepsilon_i)I_i^{\downarrow})/\tau_i\varepsilon_i+1)$, where C1 is 14387.7 μm·K, C2 is 1.19104 × 108 W·μm4·m^{-2}·sr^{-1}. We then test the urban heat island effect by constructing a model with $T_{s(UHI)} = ln(T_{s(HS)} - T_{s(v)})$, where $T_{s(UHI)}$ is the extra land surface heat in the housing estate as compared to the vegetation area nearby, $T_{s(HS)}$ is the land surface temperature of the housing estate, and $T_{s(v)}$ is the land surface temperature of the vegetation area nearby (Yu et al., 2014).

3.5.3 Sentiment Index and PCDD Data Collection

The sentiment index is constructed based on social media listening data. Via a Social listening platform (e.g. K-matrix) that uses a natural language processing approach, the unstructured components of social media can be converted to a quantitative record of the positive and negative sentiment of 55 housing estates, as well as Hong Kong's housing market in general, over the past ten years. Google Search volume data of the Chinese term for "Hong Kong housing prices" will be downloaded from Google Trends. Environmental pollutant data such as PCDD will be collected from the Environmental Protection Department. We will also visit the Land Registry Department to search for the number of years left on land leases.

3.5.4 Control Variables

Similar to other housing price modelling systems, we will include the following control variables of housing demand-side factors: (1) socioeconomic, (2) housing characteristics, (3) environmental, (4) financial factors, (5) policy, and (6) housing supply as per the literature. All data sources are listed in Table 1. Rationales are as follows:

Socioeconomic

Previous studies have revealed that places with more males and married couples are associated with higher housing prices. Furthermore, a rise in income positively affects housing demand, and an increase in the unemployment rate reduces housing demand (Lin et al., 2014). At the same time, millennials (Green & Lee, 2016) and migration (Costello et al., 2019) drive housing demand. Thus, we will include different age groups' populations, marriage status, female and male population, unemployment rate, income data from the Census and Statistics Department and net migration data from Moody's Analytics.

Financial Factors

Li and Li (2018) found that when stock prices rise, more investors are drawn to the stock market instead of the housing market. Yet, increases in the money supply will lead to a rise in housing prices. The low-interest rate environment after the global financial crisis drove up housing prices. While renminbi appreciation increased the purchasing power, it positively correlated with housing price (Li and Chau, 2016). As such, in addition to including interest rates, we will obtain (i) global money supply data from OECD, (ii) the adjusted Close Heng Seng Index from Yahoo! Finance, and (iii) the Hong Kong Real Interest Rate from the Hong Kong Monetary Authority, (iv) Exchange rate (Hong Kong to RMB) from investing.com as control variables.

Housing Characteristics

Prior research into housing prices includes house characteristics as control variables, such as the age and size of housing units. It has been found that transportation time to the central business district (CBD) for work negatively affects housing prices (Lin et al., 2014). Yang et al. (2018) found that 65 city's monthly housing prices in China are highly correlated. Thus, Shenzhen's housing prices will be obtained from Fang.com, the age and size of housing units will be obtained from the EPRC and transaction date to match financial, socioeconomic, and environmental factors, etc. Travel time will be obtained from Google Maps.

Environmental Externalities

CO, NO_2, PM2.5, PM10, SO_2, and O_3 will be obtained from the Environmental Protection Department (www.epd.gov.hk). Refer to A3 for previous findings.

Housing Supply

Restricted housing supply is a crucial driver of the uptrend in housing prices in the literature. Thus, we will include newly completed flats (the first-hand market supply) and vacancy rate (the second-hand market supply) as control variables, thereby extending the research of Li and Li (2018).

Policy

The HKSAR government introduced a Special Stamp Duty Policy in Nov 2010, a Buyers' Stamp Duty and an extension of the Special Stamp Duty in Oct 2012, and a Double Stamp Duty in February 2013 to curb increases in housing prices (Wu et al., 2017). As these increase the cost of housing purchases, we include the Stamp Duty as a control dummy variable.

- D refers to daily data, M refers to monthly data, Q refers to quarterly data, S refers to semi-annual data, and Y refers to annual data. A cubic spline interpolation will be adopted for less frequent data, and we may omit some relative less critical factors to improve prediction efficacy.
- The percentage of the data used for training, testing, and model validation will be determined by trial and error as per prediction efficacy, i.e. RMSE and MAE values.
- Data collected for this project may be used for other machine learning algorithms/econometric models in the future.
- **These units are closest to the based stations (Table 2).

Table 2 Average number of years left on land leases in Hong Kong

Estate	Years	Expired year	Estate	Year remains	Expired year
Queens Terrence	28	2047	Caribbean Coast Albany Cove	28	2047
The Merton	33	2052	Park Island	28	2047
Lei King Wan	106	2125	Bellagio	32	2051
South Horizons	21	2040	Coastal Skyline	28	2047
Heng Fa Chuen	116	2135	Tseung Kwan O Plaza	28	2047
Taikoo Shing	880	2899	Yoho Town	28	2047
Kornhill	115	2134	Residence Oasis	33	2052
Nan Fung Sun Chuen	106	2125	Golden Lion Garden	28	2047
Baguio Villa	840	2859	Sunshine City	28	2047
Island Resort	28	2047	City One Shatin	28	2047
Chi Fu Fa Yuen	32	2051	Kingswood Villas	28	2047
The Pacifica	35	2050	Metro City	28	2047
Whampoa Garden	119	2134	Garden Rivera	28	2047
Laguna City	32	2047	Dawning Views	28	2047
Amoy Gaden	32	2047	Ocean Shores	28	2047
Mei Foo Sun Chuen	32	2047	The Parcville	30	2049
Whampoa Estate	19	2034	Park Central	28	2047
Laguna Verde	32	2047	Tsuen Wan Centre	28	2047
Galaxia	32	2047	Tai Po Centre	28	2047
Royal Peninsula	32	2047	Tai Hing Gardens	28	2047
Metro Harbour View	34	2049	Tuen Mun Town Plaza	28	2047
Banyan Garden	34	2049	Tierra Verde	28	2047
Liberte	34	2049	Central Park	32	2047
Sky Tower	73	2088	Belvedere Garden	28	2047
Telford Garden	32	2047	Riviera Garden	28	2047

3.6 Contributions and Implications of the Research Framework

3.6.1 Policy and Research Implications

Social media has become important in our decision-making processes. For example, posts and comments on Facebook can influence consumers' purchasing decisions. This has given rise to social listening tools, which collect social media content and analyse sentiment using AI. From an academic perspective, we contribute to the existing literature by considering AI interpreting market sentiment through social listening. Our study is also the first of its kind to utilise remote sensing data to test the impact of the urban heat island effect on housing price predictions.

From a policy perspective, our findings will facilitate a more informed decision-making process and provide important insights for policymakers. As per social media interpreted by AI, higher housing market sentiment is associated with higher housing prices. In that case, it understands that the impact of sentiment expressed in social media is more important than controlling the housing price level to improve housing affordability among policymakers, particularly for living demand and curbing investment or speculative demand.

3.6.2 Practical Contributions

Since many valuers around the world, who are not experts in machine learning, provide valuation services using subjective human judgment in terms of housing prices, our results could help shape and improve professional practices and influence existing valuation methods and ideas. The results may also enhance valuers' awareness and understanding of AutoML, leading to innovative entrepreneurial activity in designing and delivering new automated valuation products or services.

While the impact of socioeconomic factors on housing prices is well-understood, little is known about social listening interpreted sentiment in social media among valuers. At the same time, for property developers, this study will help predict future housing prices and facilitate a more informed decision-making process in terms of land purchases and property development. The algorithms could be used as a property valuation tool for households, being aware of the role of sentiment and other factors in home purchase decision-making and optimising wealth allocation.

Acknowledgment This work described in this paper was fully supported by a grant from the Research Grants Council of the Hong Kong Special Administrative Region, China (Project No. UGC/IIDS15/E01/19).

References

Adanet (2018) AdaNet documentation. URL https://adanet.readthedocs.io/en/v0.9.0.

Besley, T., & Mueller, H. (2012). Estimating the peace dividend: The Impact of violence on house prices in Northern Ireland. *American Economic Review, 102*, 810–833.

Chen Z., Li C. & Sun W. (2020) Bitcoin price prediction using machine learning: An approach to sample dimension engineering. *Journal of Computational and Applied Mathematics* 365, 112395.

Chevapatrakul, T., Xu, Z., & Yao, K. (2019). The impact of tail risk on stock market returns: The role of market sentiment. *International Review of Economics & Finance, 59*, 289–301.

Clegg F.M., Sears M., Friesen M., Scarato T., Metzinger R., Lee Russell C., Stadtner A. & Miller A. B. (2019) Building science and radiofrequency Radiation:What makes smart and healthy buildings. *Building and Environment*, 106324.

Collins, W. J., & Margo, R. A. (2007). The economic aftermath of the 1960s riots in American cities: Evidence from property values. *The Journal of Economic History, 67*, 849–883.

Costello, G., Leishman, C., Rowley, S., & Watkins, C. (2019). Drivers of spatial change in urban housing submarkets. *The Geographical Journal, 185*, 432–446.

Ducange, P., Fazzolari, M., Petrocchi, M., & Vecchio, M. (2019). An effective decision support system for social media listening based on cross-source sentiment analysis models. *Engineering Applications of Artificial Intelligence, 78*, 71–85.

Gautier P.A. & Vuuren A. v. (2019) The effect of land lease on house prices. *Journal of Housing Economics, 46*, 101646.

Ghani, N. A., Hamid, S., Targio Hashem, I. A., & Ahmed, E. (2019). Social media big data analytics: A survey. *Computers in Human Behavior, 101*, 417–428.

González-Fernández, M., & González-Velasco, C. (2018). Can Google econometrics predict unemployment? Evidence from Spain. *Economics Letters, 170*, 42–45.

Google (2021) GCP AutoML Tables. URL https://cloud.google.com/automl-tables.

Green, R. K., & Lee, H. (2016). Age, demographics, and the demand for housing, revisited. *Regional Science and Urban Economics, 61*, 86–98.

H2o.ai (2021) Democratise AI. URL https://www.h2o.ai.

Hardell, L., Carlberg, M., & Hedendahl, L. K. (2018). Radiofrequency radiation from nearby base stations gives high levels in an apartment in Stockholm, Sweden: A case report. *Oncology Letters, 15*, 7871–7883.

Ho, W. K. O., Tang, B. S., & Wong, S. W. (2021). Predicting property prices with machine learning algorithms. *Journal of Property Research, 38*, 48–70.

Hui, E. C. M., & Wang, Z. (2014). Market sentiment in private housing market. *Habitat International, 44*, 375–385.

Hui, E. C. M., Dong, Z., Jia, S., & Lam, C. H. (2017). How does sentiment affect returns of urban housing? *Habit International, 64*, 71–84.

Jang, H. (2019). A decision support framework for robust R&D budget allocation using machine learning and optimisation. *Decision Support Systems, 121*, 1–12.

Jiang M., Liu J., Zhang L. & Liu C. (2020) An improved Stacking framework for stock index prediction by leveraging tree-based ensemble models and deep learning algorithms. *Physica A: Statistical Mechanics and its Applications* 541, 122272.

Johann, A. L., de Araújo, A. G., Delalibera, H. C., & Hirakawa, A. R. (2016). Soil moisture modeling based on stochastic behavior of forces on a no-till chisel opener. *Computers and Electronics in Agriculture, 121*, 420–428.

Khan M.A., Hernandez J.A. & Shahzad S.J.H. (2019) Time and frequency relationship between household investors' sentiment index and US industry stock returns. *Finance Research Letters* 36, 101318.

Lee, K. M., Yoo, J., Kim, S. W., Lee, J. H., & Hong, J. (2019). Autonomic machine learning platform. *International Journal of Information Management, 49*, 491–501.

Leung, T. C., & Tsang, K. P. (2013). Can anchoring and loss aversion explain the predictability of housing prices? *Pacific Economic Review, 18*, 41–59.

Li, R. Y. M., Chau, K. W., Poon, S. W., & Ho, D. C. W. (2017). Dynamic panel analysis of construction accidents in Hong Kong. *Asian Journal of Law and Economics, 8*, 20160022.

Li, R. Y. M., & Chau, K. W. (2016) *Econometric analyses of international housing markets*, Routledge, pp.1–198

Li, R. Y. M., Cheng, K. Y., & Shoaib, M. (2018). Walled buildings, sustainability, and housing prices: An artificial neural network approach. *Sustainability, 10*, 1298.

Li, R. Y. M & Li, H. C. Y. (2018). Have housing prices gone with the smelly wind? Big data analysis on landfill in Hong Kong. *Sustainability, 10*, 341.

Li, R. Y. M., Li, H. C. Y., Mak, C. K., & Tang, T. (2016). Sustainable smart home and home automation: Big data analytics approach. *International Journal of Smart Home, 10*, 177–187.

Li, R. Y. M., Tang, B., & Chau, K. W. (2019). Sustainable construction safety knowledge sharing: A partial least square-structural equation modeling and a feedforward neural network approach. *Sustainability, 11*, 5831.

Lin, W. S., Tou, J. C., Lin, S. Y., & Yeh, M. Y. (2014). Effects of socioeconomic factors on regional housing prices in the USA. *International Journal of Housing Markets and Analysis, 7*, 30–41.

Mathew, A., Sreekumar, S., Khandelwal, S., & Kumar, R. (2019). Prediction of land surface temperatures for surface urban heat island assessment over Chandigarh city using support vector regression model. *Solar Energy, 186*, 404–415.

McKee, M., & Stuckler, D. (2020). If the world fails to protect the economy, COVID-19 will damage health not just now but also in the future. *Nature Medicine, 26*, 640–642.

Mei, Y., Gao, L., Zhang, J., & Wang, J. (2020). Valuing urban air quality: A hedonic price analysis in Beijing, China. *Environmental Science and Pollution Research, 27*, 1373–1385.

Niesert, R. F., Oorschot, J. A., Veldhuisen, C. P., Brons, K., & Lange, R. J. (2019). Can Google search data help predict macroeconomic series? *International Journal of Forecasting, 36*, 1163–1172.

Orimoloye, L. O., Sung, M. C., Ma, T., & Johnson, J. E. V. (2020) Comparing the effectiveness of deep feedforward neural networks and shallow architectures for predicting stock price indices. *Expert Systems with Applications, 139*, 112828.

Regona, M., Yigitcanlar, T., Xia, B., & Li, R. Y. M. (2022). Artificial intelligent technologies in the construction industry: How are they perceived and utilized in Australia? *Journal of Open Innovation: Technology, Market, and Complexity, 8* (1), 16.

Ruan, Q., Wang, Z., Zhou, Y., & Lv, D. (2020). A new investor sentiment indicator (ISI) based on artificial intelligence: A powerful return predictor in China. *Economic Modelling, 88*, 47–58.

Sailunaz, K., & Alhajj, R. (2019) Emotion and sentiment analysis from Twitter Text. *Journal of Computational Science, 36*, 101003.

Schwarz, M., & Zimmermann, N. E. (2005). A new GLM-based method for mapping tree cover continuous fields using regional MODIS reflectance data. *Remote Sensing of Environment, 95*, 428–443.

Sekertekin, A. (2019) Validation of physical radiative transfer equation-based land surface temperature using Landsat 8 satellite imagery and SURFRAD in-situ measurements. *Journal of Atmospheric and Solar-Terrestrial Physics, 196*, 105161.

Song, L., Li, R. Y. M., & Yao, Q. (2022). An informal institutional comparative study on occupational safety knowledge sharing via French and English Tweets. *Safety Science, 147*, 105602.

Tama, B. A., & Comuzzi, M. (2019). An empirical comparison of classification techniques for next event prediction using business process event logs. *Expert Systems with Applications, 129*, 233–245.

Updyke K., Siddiqui F., Urso B., Domozych R., Brooks I. & Solomon J. (2017) Social media "listening" used to compare patient views on Enbrel to other treatments for psoriasis. *Journal of the American Academy of Dermatology, 76*, AB239.

Wang, W., Liu, K., Tang, R., & Wang, S. (2019). Remote sensing image-based analysis of the urban heat island effect in Shenzhen, China. *Physics and Chemistry of the Earth, Parts a/b/c, 110*, 168–175.

Waring J., Lindvall C. & Umeton R. (2020) Automated machine learning: Review of the state-of-the-art and opportunities for healthcare. *Artificial Intelligence in Medicine* 104, 101822.

Wiese, M., Martínez-Climent, C., & Botella-Carrubi, D. (2020). A framework for Facebook advertising effectiveness: A behavioral perspective. *Journal of Business Research, 109*, 76–87.
Wong, G. (2008). Has SARS infected the property market? Evidence from Hong Kong. *Journal of Urban Economics, 63*, 74–95.
Wu, T., Cheng, M., & Wong, K. (2017). Bayesian analysis of Hong Kong's housing price dynamics. *Pacific Economic Review, 22*, 312–331.
Wu, W., Jiang, S., Liu, R., Jin, W., & Ma, C. (2020). Economic development, demographic characteristics, road network and traffic accidents in Zhongshan, China: Gradient boosting decision tree model. *Transportmetrica a: Transport Science, 16*, 359–387.
Xu, P., Chen, Z., Wu, L., Chen, Y., Xu, D., Shen, H., Han, J., Wang, X., & Lou, X. (2019). Health risk of childhood exposure to PCDD/Fs emitted from a municipal waste incinerator in Zhejiang, China. *Science of the Total Environment, 689*, 937–944.
Yadav, D., Dansena, P., Ghosh, S.K., & Singh, P.K. (2020) A unique multilayer perceptron model (ANN) for different oxide/EG nanofluid's viscosity from the experimental study. *Physica A: Statistical Mechanics and its Applications, 549*, 124030.
Yang, J., Yu, Z., & Deng, Y. (2018). Housing price spillovers in china: A high-dimensional generalised var approach. *Regional Science and Urban Economics, 68*, 98–114.
Yang, L., Chau, K. W., & Wang, X. (2019). Are low-end housing purchasers more willing to pay for access to basic public services? evidence from China. *Research in Transportation Economics,* 76, 100734.
Yao, Q., Li, R. Y. M., Song, L., & Crabbe, J. (2021). Construction safety knowledge sharing on Twitter: A social network analysis. *Safety Science, 143*, 105411.
Yigitcanlar, T., Kankanamge, N., Regona, M., Maldonado, A., Rowan, B., Ryu, A., Desouza, K. C., Corchado, J. M., Mehmood, R., & Li, R. Y. M. (2020). Artificial Intelligence Technologies and Related Urban Planning and Development Concepts: How Are They Perceived and Utilized in Australia? *Journal of Open Innovation Technology, Marketand Complexity, 6*(4), 187.
Yu, X., Guo, X., & Wu, Z. (2014). Land surface temperature retrieval from Landsat 8 TIRS—Comparison between radiative transfer equation-based method, Split window algorithm and single channel method. *Remote Sensing, 6*, 9829–9852.
Zhai, B., & Chen, J. (2018). Development of a stacked ensemble model for forecasting and analysing daily average PM2.5 concentrations in Beijing. *China. Science of the Total Environment, 635*, 644–658.
Zhang Q., Hu W., Liu Z., & Tan J. (2020). TBM performance prediction with Bayesian optimisation and automated machine learning. *Tunnelling and Underground Space Technology, 103*, 103493.
Zheng, X., Chau, K. W., & Hui, E. C. M. (2015). Liquidity risk and cross-sectional return in the housing market. *Habitat International, 49*, 426–434.
Zhou, Z. (2018). Housing market sentiment and intervention effectiveness: Evidence from China. *Emerging Markets Review, 35*, 91–110.

Chapter 10
A Review on Sustainable Smart Homes and Home Automation in TMall, Baidu and Know the Topic: Big Data Analytics Approach

Jia-Yue Peng, Di Zhang, Ya-Wen Deng, and Rita Yi Man Li

Abstract A smart home provides residents with medical care, life assistance, security, and environmental sustainability. At the same time, automation technology has become the core part of innovative home technology that includes the Internet of Things, artificial intelligence and blockchain. This research uses a big data approach to extract and analyze the keywords related to the smart homes on various websites such as Baidu, TMall, Know the Topic and understands the smart home market in China. The results reveal that Beijing and Shenzhen have the most significant number of searches for the smart homes. The top three most popular smart appliances include smart toilets, speakers, and television. People aged 30–39 are the group with the largest number of individuals searching for smart homes.

Keywords Smart home · Automation technology · Big data analysis

1 Introduction

1.1 Smart Home Concept

In the 1980s, the concept of a smart home began to appear in the Western world. When the twenty-first century comes, smart home technologies have entered China, but domestic development remains slow. The rapid development of science and technology in China promote smart home products development. This involves smart

Equal contributions among the first, the second and the third authors.

J.-Y. Peng · D. Zhang
Sichuan City Vocational College, Chengdu 610101, Sichuan, China

Y.-W. Deng
IFLYTEK South China Co., Ltd, Guangzhou 511457, Guangdong, China

R. Y. M. Li (✉)
Hong Kong Shue Yan University, Hong Kong, China

R. T. M. Li et al. (eds.), *Current State of Art in Artificial Intelligence and Ubiquitous Cities*, https://doi.org/10.1007/978-981-19-0737-1_10

security, home appliance control, lighting control, and network services. Automation technology is at the core of the smart home, and all smart services are built based on automation.

Smart homes provide residents with a medium for socializing and reshaping the environment. Smart homes with diversified sensor networks and intelligent technologies provide support and response management to master the scope of social, healthcare, independent living, security, safety, and environmental sustainability benefits (Abdul Rehman Javed et al., 2021). Smart homes support automation sustainability to encourage smart cities, communities, and high-tech-driven solutions (Aleksandra (Ola) Michalec et al., 2019).

From the perspective of energy-efficient buildings and comfortable living, smart homes have aroused more and more interest. The availability of many ICT and energy technologies allows residents to integrate technology and services into their houses quickly. A smart home is "a living environment with information and communication technology that provides suitable functions for residents' convenience, safety, entertainment and comfort needs." The central concept of a smart home is to use well-designed management functions to provide residents with the best ability to control house maintenance (Wonyoung Choi et al., 2021).

Research on smart houses has been going on for decades. Each house will have a global network of interconnected devices built into it, and these devices will be connected to promote simultaneous connections around the world. Typical electronic devices in the room, such as temperature sensors, light switches, and thermostats, help manage the atmosphere and help regulate the temperature. This interaction is more beneficial, easier to use, and encourages residents to continue their daily work. Some studies have found that if smart device interaction is carried out in a more "natural" environment, the environment will be more straightforward, practical, and functional (Radha, 2021).

The smart home market is expected to grow to 174 billion by 2025 (Marr, 2020). This popularity stems from many innovative smart home devices that can make our lives more comfortable, safe and convenient-such as devices that monitor personal health (such as monitoring sleep, tracking heart activity), and improve safety (such as reminding caregivers). Some equipment can learn and adapt to the owner's preferences and behaviours. The smart home has been developed by utilising engineering progress, processing speed improvement, robust network, data analysis, and artificial intelligence (DeFranco, 2021).

1.2 Smart Home Automation

Domestic intelligent automation technology is being promoted in China, and the popularity of smart homes is increasing. In recent years, some real estate companies have often used "smart home" as their selling point and provide residents with more intelligent, automatic and convenient life services to improve people's living standards continuously.

The Internet of Things era is brought about by advanced science and technology and smart home automation technology no longer exists in our minds and movies. As the technology continues to improve, smart household products that are fully functional, economical, practical, and easy to use have also begun to appear. Such products quickly gain popularity in the domestic market and become essential items in people's daily lives. The main reason is that the changes made by many home furnishing manufacturers in their corporate operating methods and business mentality have caused a considerable change in the concept of the entire market.

To a certain extent, home automation is equivalent to a smart home, that is, the realization of home intelligence, making family life more convenient and convenient, and more enjoy the changes brought by high technology. However, there are still fundamental differences between home automation and smart home.

Home automation mainly uses electronic technology to control household appliances and security equipment in a unified manner through a central processing unit, making family life more convenient and comfortable and turning the manual into automatic operation. However, in addition to automation, smart homes also include information and multimedia platforms. The entire system has a "brain" like a robot with intelligence. For example, the air conditioner can be turned on through a mobile phone or telephone without manual remote control. This is to achieve automation, while smart homes can determine the specific switching time of the air conditioner in the home through the judgment of the external environment, such as air humidity and temperature, etc and reach the most comfortable temperature.

Regarding smart home technology, Sovacool and Furszyfer Del Rio (2020) summarized ten definitions of smart home, from the integration of different services in the home environment to a broader technology set that provides customized services to users.

The most prominent examples of innovative home technology include smart speakers Amazon Alexa and Echo and Google Home, whose virtual assistants have voice-controlled services such as streaming music, searching the internet, ordering products, and checking the weather. Nest sells automatic and digital thermostats, indoor and outdoor security cameras, smoke and carbon monoxide alarms, security systems, energy monitors and video systems.

However, previous studies have shown that smart home technology must have the ability to digitally connect with different devices and information sources to provide users with more customized services (Dylan D. Furszyfer Del Rio et al., 2020). Based on this, smart home technology should potentially affect family life by providing personal health feedback, enhancing energy demand and control, and even ensuring family safety and user well-being.

Smart home technology is not limited to the home entity but also connects them to the outside world through remote control and connection to cloud-based services. In the Internet of Things, technologies and services such as cloud computing services, social networks, and smart home technologies collect and share data about how devices are used and the environment.

On the other hand, automation and robotics are expected to replace many previously manual tasks. This digital revolution will bring about social changes that may have a wide-ranging, profound and possibly lasting impact and will also enter our family space. For example, in the United Kingdom, a survey conducted by The

Guardian (2019) revealed how to spend millions of pounds on developing a new generation of welfare robots to replace humans. This process gave rise to a whole new term, such as "virtual labour.", "enhanced decision-making", and "robot process automation", and so on. As a result, people are increasingly worried that the human factor in the welfare state is declining. Therefore, discrimination and inequality against people who have less technical knowledge or cannot access the internet may worsen.

Islam et al. proposed a robust architecture for short-distance and remote home automation using multiple communication technologies (LoRaWAN, server-based LoRa gateway, and Bluetooth connection). The integrated system effectively controls different household appliances and maintains intelligent management among all electronic components (Rahabul Islam et al., 2021).

Mukherjee et al. studied unified smart home resource access and identity verification using blockchain technology. By combining the power of blockchain technology and public-key cryptography, the security authentication and identification of smart home personnel and equipment can be ensured. This work avoids repeated authentication of access to the same service and communicates within a limited delay. Most essential security goals such as authentication, authorization, confidentiality and integrity are met (Arnab Mukherjee et al., 2021).

With the development of artificial intelligence, smart homes are playing an increasingly important role in daily life. Since new objects may often appear in the home and collecting enough training samples is sometimes tricky, the task of few-shot recognition is essential and practical in smart home scenes. Chang et al. proposed a migration learning algorithm based on multi-core learning for multi-class migration incremental learning based on multi-core learning (Chang et al., 2020).

1.3 The Bottleneck of the Development of Home Automation

The development of home automation in China is relatively slow. It mainly stays in the experience hall and some pilot buildings, which is different from providing a comfortable and prosperous living environment, a convenient and flexible lifestyle, and efficient and reliable working mode. Some projects that are highly anticipated by users and technically supported have not been commercially available on a large scale. The main reasons are as follows:

(1) Industry standards are difficult to integrate

Due to historical reasons and economic interests, the two major domestic standards, e Jiajia and IGRS, currently represent home appliance manufacturers and IT manufacturers. E Jiajia has apparent advantages in intelligent home appliances, while IGRS has certain advantages in productizing device interconnection. Nevertheless, as 3C integration matures, home automation has become the most critical battlefield between the two. Cross-industry and sectoral integration and industrial chain integration are imminent.

(2) High degree of technical specialization

Unlike the plug-and-play of computer products and home appliances, the current home automation products are an overall system that requires professionals to install, debug, and use training. Users cannot experience the fun by using it, reducing the willingness to use and purchase. This reduces the value of products, raises prices and usage thresholds, and restricts the development of the industry. At the same time, product specialization also means that R&D investment is large, technology sharing is difficult, each manufacturer has their own standards. It does not pay enough attention to products, long market cultivation period, long time for customer awareness directly reflect slow progress in industrial scale. To strengthen the practicality of technology and increase the popularity of products, the priority is to solve the complexity of the technology so that more practitioners can enter the industry, reduce costs, and promote the process of marketization.

(3) Slow progress in product intelligence

A smart home involves many disciplines such as computer, communication, automation, electronics and so on. Many of the modern enterprises are only brilliant in some areas. At present, domestic products related to the home automation industry abound, but the level of intelligence is low. Its products and services are only a supplement to humans to realize some mechanical functions, rather than the main force to change the lifestyle. This is caused by the failure to carry out a reasonable division of labour, and it is also the main reason why home automation cannot be promoted in real estate.

(4) Security issues

Since smart home systems are comprised of components that the internet can control, these systems can quickly become the primary source of threats to our private lives due to the security vulnerabilities of these systems. Yassein and others investigated security issues and solutions that compromise the integrity of intelligent home systems. It is constructed by dividing the intelligent home system into its main components and then investigating each component individually, defining and discussing each component's main security issues and challenges (Muneer Bani Yassein et al., 2019).

2 Big Data Analysis Method

Big data analysis refers to the analysis of large-scale data. It can be big due to the challenges which may arise in an experimental setting and be defined as a huge structured and unstructured data from different sources (Li & Li, 2018). Big data can be summarized as 5 Vs, with large data volume (Volume), fast speed (Velocity), multiple types (Variety), value (Value), and authenticity (Veracity).

Big data is the hottest vocabulary of the IT industry. The use of data warehouse, data security, data analysis, data mining and so on around the commercial value of big data has gradually become the focus of profit sought after by industry professionals. With the advent of the era of big data, big data analysis has also emerged.

Everyone is a data contributor. The average person in the world carries about 200 GB of data. As an emerging field around people and devices, smart home is self-evident that big data contribute to each other. The data generated by the smart home system includes a wide range of data, including the data of hardware sensors, the data running status of the hardware itself, and the data of the interaction between the user and the hardware, and the data generated by the user through the App and other clients. The user's usage habits and life scene data have led to a considerable accumulation of smart home data. Innovative home companies focus on business-driven early-stage, and some of the following data must be collected. It is recommended to adopt a distributed large-scale cloud storage architecture to meet the inevitable trend of rapid development and innovation needs of enterprises in the future.

Smart home big data collection includes App usage, fault self-diagnosis information, service operation information, user portraits, device usage status, user usage behaviour, App interaction behaviour, user information data, device function information, user information, and equipment function information, equipment logs, APP logs, sub-equipment parameters and operating status and other data.

A smart home is the entry point of multi-field integration, the support point of social family management, and the new bright spot of people's livelihood services. The Internet of Things produces big data, and big data support smart homes. From smart homes to data to intelligence, the whole process from perception to cognition is constituted. Big data is the commanding height of smart hardware competition. It can help hardware manufacturers dig out users' equipment usage behaviour to understand their users and optimize product strategies and market strategies. It can also learn from users, build user portraits, provide personalized intelligent experiences for different users, provide discounts, and increase secondary sales. We can also do the analysis of equipment activity status and failure rate to offer advice to the product and hardware designers regarding how to do iterative layers and how to upgrade, including knowing what function users like to use, when users like to use this function, and knowing what follows and which regions are the focus of the marketing strategy. The use of big data can provide much value, which requires continuous exploration in the IoT field.

To obtain massive amounts of data, devices must be connected to the cloud, and the smart home field will face hundreds of billions or even trillions of devices. If cloud security deployment is not mature, blindly connecting these devices into the cloud will lead to unimaginable disasters. Therefore, the data premise is cloud security, so the method of cloud computing is significant.

Cloud computing and big data are two sides of the same coin. Cloud computing is the IT foundation of big data, and it is a killer application of cloud computing. Cloud computing is the driving force for the growth of big data. On the other hand, more and more data need to be processed by cloud computing. Therefore, cloud computing and big data are complementary and are more vividly reflected in the smart home industry.

In the future, while storing, analyzing and accurately extracting big data through cloud computing, deep learning and in-depth data mining are required to simulate user behaviour and achieve a more thoughtful and intelligent experience.

Many private companies and state-owned companies have significant interests in massive data collection. After following the trend to build a big data platform, the result achieved is PB-level storage capacity and second-level processing capacity. It may not be connected to the public cloud to collect a large amount of data, and in the end, it will only be stored first and used when needed. This kind of thinking is not necessary. With the continuous improvement of big data technology, its flexibility, transparency, and robustness will continue to improve, and adequate storage, management, and maintenance forms are constantly changing. Feasibility and practicality are the essences of big data.

3 Research Results

3.1 Purpose

This book chapter mainly throws light on the following points:

- The distribution of various smart products
- City distribution, monthly search volume, age distribution, and gender distribution of users searching for "smart home" keywords
- Search for topics related to "smart home" on the internet and conduct word frequency analysis
- Word frequency analysis of user reviews of representative products.

3.2 Data Sources

This data mainly comes from the following websites:

- Baidu Index
- Know the topic (reptiles)
- Tmall comments (crawler).

3.3 Analysis Process

(1) Search the distribution of various smart products

Figure 1 shows the relationship between most of the smart home products on the market in the past 30 days and the search volume of users on Baidu. This table shows that the search volume of "smart toilet" is the highest, followed by smart speakers and smart TVs, and the others are relatively low. The smart refrigerator has the lowest search volume, with 0 search volume.

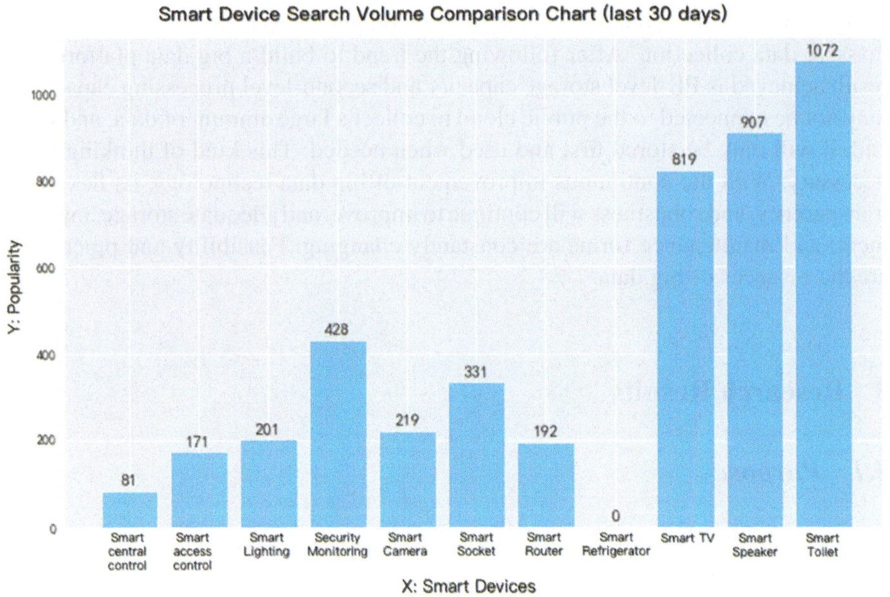

Fig. 1 Search distribution of various smart products (nearly 30 days)

(2) Analysis of indexes of searching for "smart home" keywords (nearly 30 days)

Figure 2 shows the distribution of users who searched for "smart home" keywords in the past 30 days, with the top four first-tier cities in Beijing, Shanghai, Guangzhou and Shenzhen. There is a certain relationship between search volume and urban development, and cities with large search volumes generally have better economic development. The reason may be that first-tier cities have high population density, a good living environment, and high per capita income which supports them to buy smart home appliances.

(3) Time distribution of users searching for "smart home" keywords (2017–2018)

It can be seen from the above chart that the search volume before and at the end of the two years was smaller than that other months. July 2017 and May 2018 have the highest user search volume in the past 2 years (Fig. 3).

Regarding the reason for the two abnormal peaks, through inquiries, in July 2017, it is likely to be affected by the "Fifth Shanghai Smart Home Exhibition", and in May 2018, it may be affected by the "third session". The "China Smart Home International Summit" impact was relatively small before and at the end of the year. "Smart home" is not a topic of much concern at this time.

(4) The age distribution of users who search for the keyword "smart home"

It can be seen from Fig. 4 that most of the users searching for the keyword "smart home" is 30–39 years old, followed by 40–49 years old; users are less than 19 years

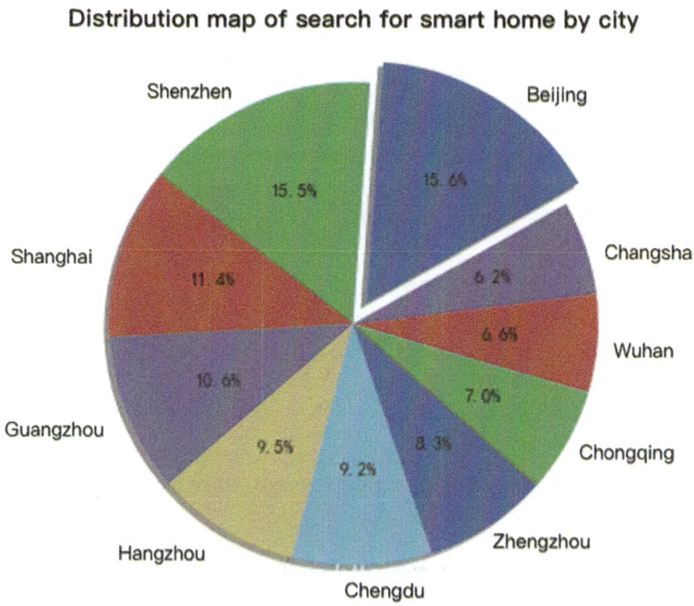

Fig. 2 Analysis of the location of users who search for "smart home" keywords (some areas with small data volumes are not shown)

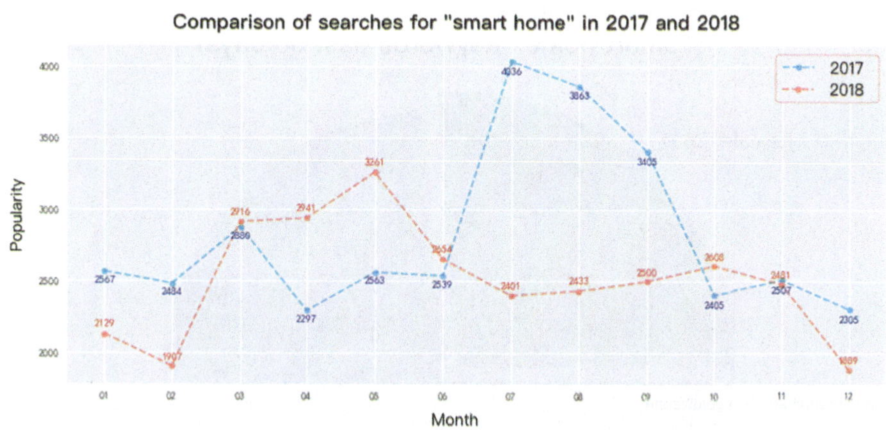

Fig. 3 Time distribution of users searching for the keyword "smart home" from 2017 to 2018

old and less than 50 years old. The reason may be that between the ages of 30 and 49, users in this age group tend to have stable work and life and higher incomes; among those over 50, their they may have no idea on "smart home" concept.

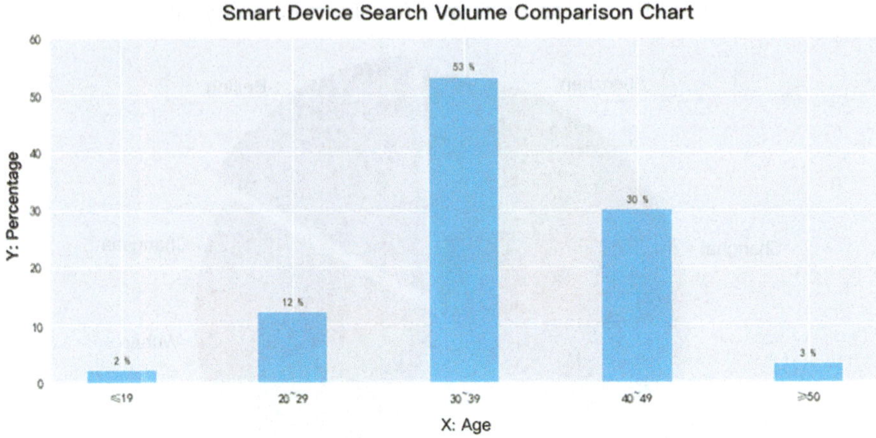

Fig. 4 Age distribution of users searching for the keyword "smart home"

(5) Gender distribution of users who search for "smart home" keywords

It can be seen from Fig. 5 that most of the users who search for "smart home" are "male", accounting for 3/4 of the total search.

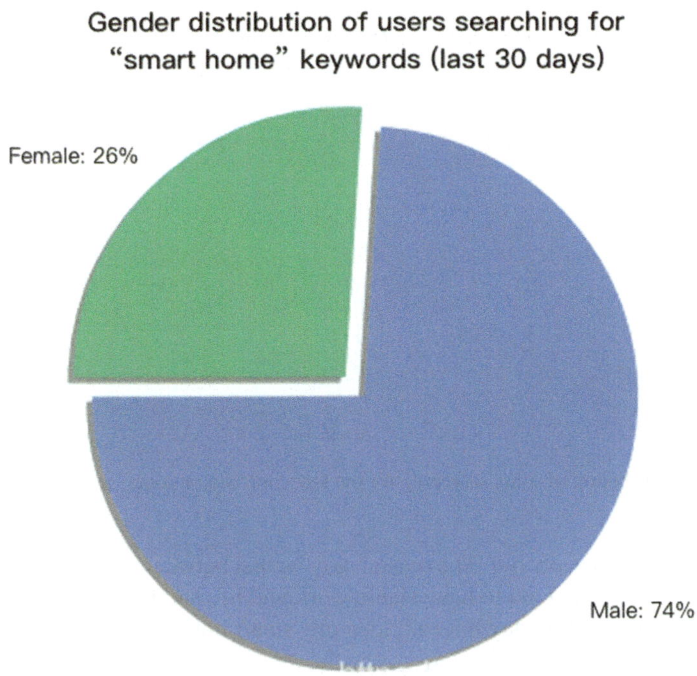

Fig. 5 Gender distribution of users searching for "smart home" keywords

Fig. 6 Word frequency analysis of Zhihu "smart home" related topics

(6) Zhihu word frequency analysis on topics related to "smart home"

After removing the irrelevant words "system", "product", and "home", it can be seen from Fig. 6 that "lighting" has the highest appearance rate in the topic of "smart home", followed by "technology", "Xiaomi" and "Words such as "industry" and "program".

Smart lighting is a popular topic and the frequent mentioned word "lighting" is related to "smart lighting system" and "lighting control".

"Technology" ranks second in the topic of Zhihu, which shows that Zhihu users are more concerned about the technology used in intelligent products;

"Xiaomi", the word's appearance rate ranked third, indicating that Zhihu users have a relatively high discussion of "Xiaomi Smart Home", and Xiaomi has a certain influence in the "smart home" market. Let's take the more popular Xiaomi camera for Xiaomi smart home products as an example.

(7) Word frequency analysis of Xiaomi camera-Tmall comments

It can be seen from Fig. 7 that "quality" and "picture" appear most frequently in Tmall comments, followed by "effect", "image", and "clarity".

The word "Problem" appears in the comments. Some mentioned that "I have encountered problems in every possible way", "There is no problem with the connection", "There is no problem for the time being" and others.

Buyers and users mainly focus on the quality of this type of product. They pay more attention to functions such as "night vision".

Fig. 7 Word frequency analysis of Xiaomi camera-Tmall comments

4 Conclusion

Automation technology has gradually been used widely in smart homes, which has greatly improved the overall level of smart homes. More importantly, it has greatly improved people's living standards, providing people with convenient and fast life services, and saved People's time. Special features of automated smart homes, such as convenience, safety, and comfort, meet people's new demands for modern homes and make more and more people yearn for smart home life. Smart home automation technology in China will continue to develop, so that automated smart homes can truly enter millions of households.

Acknowledgments This work described in this paper was fully supported by a grant from the Research Grants Council of the Hong Kong Special Administrative Region, China (Project No. UGC/IIDS15/E01/19)

References

Abdul Rehman Javed, L. G. F., Farhan, A. A., Abbas, S., Srivastava, G., Parizi, R. M., & Khan, M. S. (2021). Automated cognitive health assessment in smart homes using machine learning. *Sustainable Cities and Society, 65*.

Aleksandra (Ola) Michalec, E. H., James Longhurst. (2019). Building smart cities, the just way. A critical review of "smart" and "just" initiatives in Bristol, UK. *Sustainable Cities and Society, 47*.

Arnab Mukherjee, M. B., Pujari, C., Tiwari, S., Nayar, A., & Payyavula, S. R. (2021). Unified smart home resource access along with authentication using Blockchain technology. *Global Transitions Proceedings, 2*(1), 29–34.

Chang, S. C., & Zhao, C. H. (2020). Multiple kernel based transfer learning for the few-shot recognition task in smart home scene. *IFAC-PapersOnLine, 53*(2), 17101–17106.

Dylan D. Furszyfer Del Rio, B. K. S., Bergman, N., & Makuch, K. E. (2020). Critically reviewing smart home technology applications and business models in Europe. *Energy Policy*.

Dylan D. Furszyfer Del Rio, B. K. S., & Griffiths, S. (2021). Culture, energy and climate sustainability, and smart home technologies: A mixed methods comparison of four countries. *Energy and Climate Change, 2*.

Joanna F. DeFranco M. K. (2021). Smart home research themes: An analysis and taxonomy. *Procedia Computer Science, 185*, 91–100.

Li, R. Y. M., & Li, H. C. Y. (2018). Have Housing Prices Gone with the Smelly Wind? Big Data Analysis on Landfill in Hong Kong. *Sustainability , 10*, 341.

Marr, B. (2020). The 5 biggest smart home trends in 2020. https://www.forbes.com/sites/bernardmarr/2020/01/13/the-5-biggest-smart-home-trends-in-2020/#7724bf96389b.

Muneer Bani Yassein, I. H., Shatnawi, F., Mardini, W., & Khamayseh, Y. (2019). Smart home is not smart enough to protect you—Protocols, challenges and open issues. *Procedia Computer Science, 160*, 134–141.

Radha, R. K. (2021). Flexible smart home design: Case study to design future smart home prototypes. *Ain Shams Engineering Journal*.

Rahabul Islam, M. W. R., Rubaiat, R., Hasan, Md. M., Reza, Md. M., & Rahman, M. M. (2021). LoRa and server-based home automation using the internet of things (IoT). *Journal of King Saud University—Computer and Information Sciences*.

Wonyoung Choi, J. K., Lee, S., & Park, E. (2021). Smart home and internet of things: A bibliometric study. *Journal of Cleaner Production, 301*.

Chapter 11
State of the Art Research in Artificial Intelligence and Ubiquitous City

Rita Yi Man Li, Daniel Chi Wing Ho, and Kwong Wing Chau

Abstract AI development has affected many industries over the last decade. For example, EmBot uses AI to search legal precedents in patent disputes and identify cancer in magnetic resonance images. In addition to AI, there is rapid development in ubiquitous technology for u-city. Ubiquitous refers to the ability to exist everywhere and be everywhere simultaneously. It has been used in computer sciences in 'ubiquitous computing' and elaborated with 'ubiquitous community' as the human presence and activity in physical and digital spaces. Given the rapid development in AI and u-city, we invited top professors in these two areas to share their thoughts and this chapter summarises their viewpoints: (1) keep abreast of the latest development trends, prospects and challenges in AI and u-cities; (2) identify various AI and ubiquitous tools that can be used in research projects; and (3) review the current state of art applications of AI for u-city.

Keywords Artificial intelligence · U-city · Smart cities

1 Background of Our Inter-Institutional Development Scheme Project

This book is a byproduct of the research grant awarded from the Research Grants Council, Inter-Institutional Development Scheme project. We have invited top scholars who researched artificial intelligence and ubiquitous city. In the following section, we will briefly introduce the theoretical background, aims and details of the project, followed by the speakers' contents of speeches.

R. Y. M. Li (✉)
Sustainable Real Estate Research Center, Hong Kong Shue Yan University, Hong Kong, China

D. C. W. Ho
Department of Environment, Technological and Higher Education Institute of Hong Kong, Hong Kong, China

K. W. Chau
Department of Real Estate and Construction, The University of Hong Kong, Hong Kong, China

R. T. M. Li et al. (eds.), *Current State of Art in Artificial Intelligence and Ubiquitous Cities*, https://doi.org/10.1007/978-981-19-0737-1_11

Andrew Ng has compared the transformative power of AI to that of electricity: "just as electricity transformed almost everything 100 years ago, today I have a hard time thinking of an industry that I don't think AI will transform in the next several years." Though his expectations verge on idealism, AI development has genuinely affected several industries over the last decade. For example, AI techniques are used to play games such as Go and have attracted public attention. EmBot used AI to accomplish tasks such as searching for legal precedents in patent disputes and checking for cancer in magnetic resonance images (Peha, 2019; Wall, 2018).

The sudden rise and exponential development of AI have progressed in finance, real estate, and construction. Netflix has also utilised AI in its movie recommendations, Amazon's voice assistant Alexa, and Google's mammoth search engine. This has directly impacted these companies' growth and praise—and vilification. Staying abreast of the latest developments in the field will give academics an advantage in searching for the latest techniques in 2020 to conduct relevant research and improve teaching methods (Wall, 2018).

In addition to AI, which is quite clearly picking up steam, ubiquitous technology for u-city is also noteworthy. Derived from the Latin word 'ubique', which means 'everywhere, ubiquitous is the ability to exist everywhere and be everywhere simultaneously'. It has been used in computer sciences in 'ubiquitous computing' and elaborated with 'ubiquitous community' as the human presence and activity in physical and digital spaces (Alanne & Cao, 2019).

UbiCom, which marries the physical world (human and social environments) with the world of the internet, data and mobile devices (Ghaemi Rad et al., 2018), involving ambient intelligence, has begun to influence health-monitoring transportation, hospitals, education, as well as the workplace. The wide adoption of UbiCom in a "ubiquitous city" like Korea include pressure-sensitive floors in homes for the elderly that can detect when someone falls and immediately call for help (Serrano & Botia, 2013). Small smart-home devices such as the Amazon Echo or the Echo Dot and robots with AI linked with other smart home devices, sensors and actuators are embedded in the physical environment. The most common places for such devices are parking lots, but even regular smartphones come with them now. These ubiquitous devices have begun to permeate our lives on a scale hitherto unimagined.

An overarching application of AI is to develop solutions for u-cities built on convergent systems and provide solutions to the problems of sustainability and pollution, i.e., issues that afflict most metropolitan cities. Ubiquitous cities are often considered advanced intelligent cities, though smart cities themselves are part of ubiquitous cities (Ghaemi Rad et al., 2018). Ubiquitous information communication technologies assist in sustaining cities and managing them more efficiently, which ultimately improves citizens' quality of life (Lam & Ma, 2018).

Smart cities usually entail each citizen carrying intelligent devices. A u-city, however, has sensors embedded at every turn so that anyone with a device can use its full potential at any time and place within the city (Ghaemi Rad et al., 2018). Smart cities also allow gathering massive amounts of data; however, it is rarely helpful and goes to waste. Hence, the dynamic nature of a ubiquitous city requires better data aggregation to optimise the city operations. For example, the shift of deep learning towards semi-supervision enhances the cognitive abilities of many smart city services by providing several use cases (Mohammadi & Al-Fuqaha, 2018). The interconnection of medical devices via the Internet of Health Things allows data from different sources to be combined, enhances patient health status diagnosis and identifies the best solutions (Da Costa et al., 2018).

In our IIDS AI and ubiquitous city project, experts will be invited to share their knowledge with local academics in three areas:

The *first area* focuses on the latest trends, developments and research topics in AI and u-cities. The *second area* throws light on the current state of the art AI technology that can be used in research projects. Scholars and experts in the programmes will be invited to discuss and demonstrate various AI and ubiquitous tools to enhance academics' research capability, hence existing research methods. For the *third area*, renowned researchers were invited to share their experiences of applying AI for u-cities research. Several successful and potential projects were given to encourage academics to use these tools in their research.

Headlined by renowned scholars and leading experts in the field, this project aims to advance the understanding of AI and u-city, bringing together local academics to discuss their ideas with international scholars. As the event was opened to industry practitioners, it inspired research agendas, introduced new methods and tools for research, and encouraged better research with practical applications. It offered a golden chance for academics to develop their research capability and enhance their field knowledge. Thus, it will bring long-lasting benefits to academia.

By the end of the project, participating academics should be able to:

1. keep abreast of the latest development trends, prospects and challenges in AI and u-cities;
2. identify various AI and ubiquitous tools that can be used in research projects;
3. review the current state of art applications of AI for u-city.

The research project is summarised in Table 1:

Table 1 Summary of the research project details

Project information	Details
Project Title (funded by Research Grant Council—2019/2020 Inter-Institutional Development Scheme)	State of the art research in artificial intelligence and ubiquitous city
Principal Investigator	Dr Rita Yi Man LI (Associate Professor, Department of Economics and Finance, Director, Sustainable Real Estate Research Center)
Co-Investigators	Chair Professor Kwong Wing CHAU, the University of Hong Kong; Professor Daniel Chi Wing HO, Technological and Higher Education Institute of Hong Kong

2 Contents of the Seminars and Workshops

This section will introduce some of the professors who spoke for our seminars and workshops.

2.1 Virginia Dignum

Professor Virginia Dignum is a professor of Social and Ethical Artificial Intelligence at Umeå University, Sweden and the scientific director of WASP-HS, the Wallenberg Program on Humanities and Society for AI, Autonomous Systems and Software. She is a Fellow of the European Artificial Intelligence Association and author of "Responsible Artificial Intelligence: developing and using AI in a responsible way", published in 2019.

She received her PhD in Artificial Intelligence from Utrecht University, and she was awarded the prestigious Veni grant by the Dutch Organization for Scientific Research in 2006. She is a well-known speaker on the social and ethical impacts of Artificial Intelligence, a member of the reviewing boards for all major journals and conferences in AI and has published over 180 peer-reviewed papers.

Professor Dignum delivered a presentation on "Responsible Artificial Intelligence". She considered AI as a technology of pattern matching, the next step in problem-solving digitisation, a field of model intelligence to understand intelligence. AI is more than just algorithm or machine learning, It can identify patterns in data (images, text and video) and extrapolate patterns and take actions based on the pattern.

However, there is a limit of AI now. AI cannot do common sense reasoning, which means that AI cannot understand context or meaning. AI cannot learn from a few examples (it requires many data for training). It cannot learn general concepts nor combine learning and reasoning.

For responsible AI, it shall be ethical, lawful, reliable, and beneficial. While responsible AI recognises that AI systems are artefacts with purpose, one of the dilemmas of automated decision-making is accuracy versus explainability. For instance, it may achieve 95% accuracy with no explanatory power. Alternatively, an AI technology attains 80% accuracy but can explain. Automated manufacturing counts as another dilemma. It relates to social and environmental wellbeing versus individual profits.

Recently, there have been many initiatives for ethical AI concerning strategies, policies and declarations. The challenge is, an endorsement is not compliance. AI should make decisions morally, socially and legally acceptable. "Design" should take a tremendous responsibility. In design, we have to ensure that the development processes take AI's ethical and social implications into account. By design, integration of ethical reasoning abilities as part of the behaviour of artificial autonomous

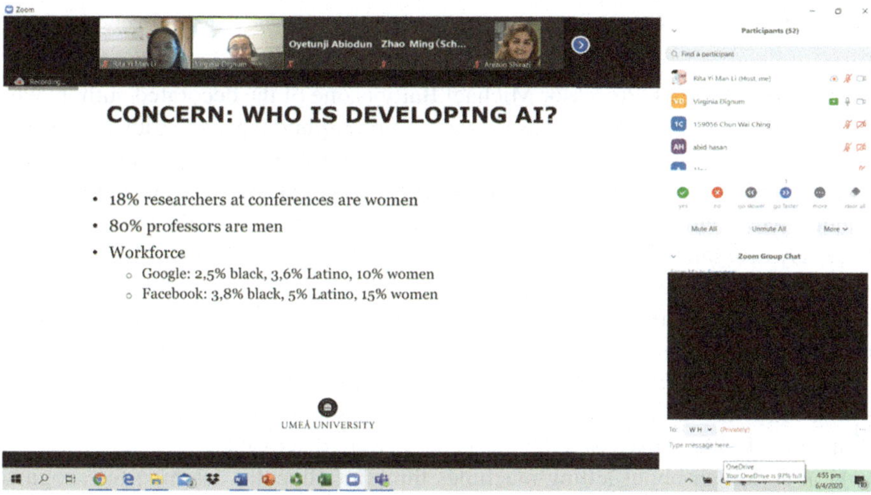

Fig. 1 Concern: who is developing AI?

systems shall be enforced. To ensure regulation and certification mechanisms, there shall be research integrity of stakeholders (researchers, developers, manufacturers …) and institutions. Besides, many of the AI developers are men. There is a concern about whether AI thinks may have a bias regarding gender-related issues (Fig. 1).

In principle, AI needs accountability, responsibility and transparency to do the things right. AI systems shall take decisions that have ethical grounds and consequences. In short, trustworthy AI has to:

- Empower and protect humans and society.
- Take up a tailored approach to the AI market.
- Secure a single European market for responsible AI.
- Enable AI ecosystems through sectoral multi-stakeholders alliances.
- Foster the European data economy.
- Exploit the multi-faceted role of the public sector.
- Strengthen and unite Europe's research capabilities.
- Nurture education.
- Adopt a risk-based governance approach to AI and ensure an appropriate regulatory framework.
- Stimulate an open and lucrative investment environment.
- Embrace a holistic way of thinking.

Her recorded video can be found in https://www.youtube.com/watch?v=s-3I21rxdSE&t=110s and https://www.youtube.com/watch?v=Nu5GKOk62PU&t=16s.

2.2 Michael Batty

As of March 14, 2020, Professor Michael Batty is one of the decorated authors who attained 38,675 Google Scholar citations on urban planning and cities topics. His research focuses on the Geographic Information System (GIS) processes, volunteered geographic information (VGI), Geographical AI etc. His current research includes AI geography-characterised big data analysis, and evaluation should be effectively designed while considering geographic mechanisms and rules, as well as the geographic knowledge inclusion.

He presented "Artificial Intelligence and Smart Cities" (Fig. 2). In the talk, he mentioned a short history of computer models of cities, the emergence of the smart city, high and low-frequency cities, big data, real-time streaming, digital twins, turning hardware to software and vice versa, examples of four different types of AI in cities. He also discussed the limits to AI were discussed.

AI is no longer about letting machines think intelligently, learning to recognise intelligence in data. A smart city is a city that generates real-time data, and we can use this data to make the city better. It can be classified as a high-frequency city or a low-frequency city. Such cities represent how we capture and use big data to improve sustainability and the quality of life.

Information (Data) is a key for smart cities. Data generated from different sources complement each other. Data is enormous in terms of its volume, and it usually implies the number of locations, their attributes and interactions. He shows a software called Quant (Fig. 3) developed by his team to analyse city data.

The idea of a digital twin is a simulation of the real thing. First models depend on the media of the original model, then to build a "digital" model same as the original

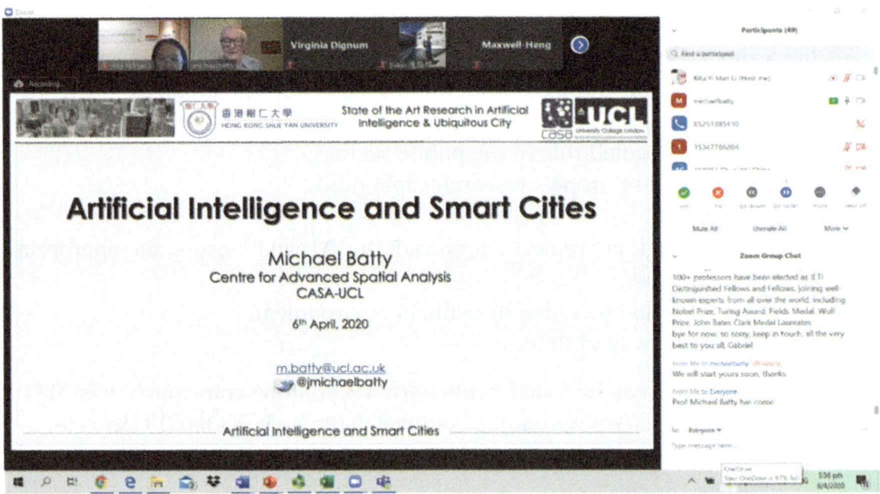

Fig. 2 Professor Michael Batty presented "Artificial Intelligence and Smart Cities"

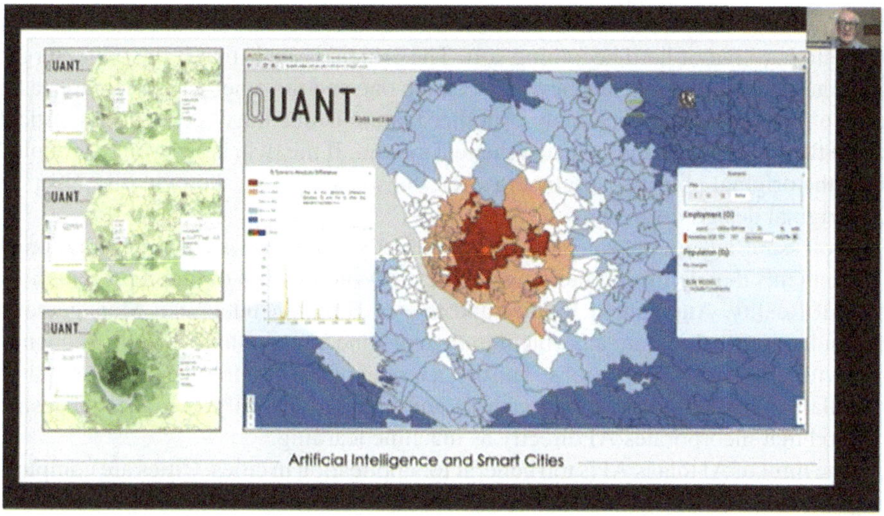

Fig. 3 Professor Michael Batty demonstrates the software which was developed by his team for analysing city-data

model is the prospect of building a twin. However, the questions lay on whether such a digital twin can predict the future? Whether the digital twin are reactive or predictive? Furthermore, how close would the prediction be to the real system? Would it transfer information to the actual model? He shows the digital twin UCL in his talk (Fig. 4).

For hardware and software transformation, building information models are classic in this regard. It is chaining architectural design and the management of

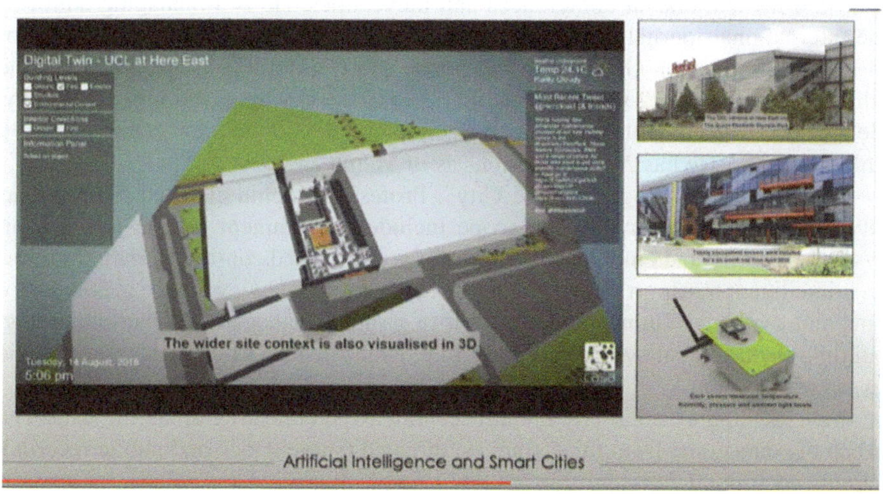

Fig. 4 Professor Micahel Batty highlights digital twin in his talk

buildings. This extends to the idea of smart city that it is not just about embedding soft into hard but of hard becoming soft. The critical issue is that if the twin merges with the artefact, then the latency between the one and the other is problematic, as it can no longer function separately. Another feature is that the dependence is critical to the function of both the twin and the real system. If the twin is so strongly coupled with the original object or artefact or even the social system, the twin disappears if the original design disappears.

Professor Batty quotes four examples are: firstly, using a physical representation of the elements that define the buildings in the city—building in digital representation, Virtual Reality, Augmented Reality and thence Building Integrated Management. Secondly, traffic flows are a problem in demand and supply: there is high-frequency city with long-term implications. Thirdly, a model of the low-frequency city requires stimulation for long-term urban growth and change. Fourthly, a model or design method that incorporates AI directly as machine learning.

The limit of AI that is AI is too general for application in cities. Cities are complex, they evolve from the bottom up, and AI helps automate relatively mundane tasks. In the future, AI can probably make an enormous impact on natural systems and the physiology of human systems.

His presentation can be found in https://www.youtube.com/watch?v=s-3I21rxdSE&t=13389s.

2.3 *Rob Kitchin*

Professor Rob Kitchin has published widely in social sciences, including 28 authored/edited books and over 180 articles and book chapters, and has delivered over 230 invited talks at conferences and universities. He is a managing editor of the international journal, Dialogues in Human Geography. He has been an editor of Progress in Human Geography and Social and Cultural Geography and was the editor-in-chief of the 12 volumes, International Encyclopedia of Human Geography. He has successfully written or been a principal investigator on forty grants with €36 million in total. He has 25,666 citations as of March 24 2020.

In the talk "The Programmable City", Professor Kitchin clarified what "smart city systems" shall contain. The scope includes government (e.g. E-government system), security and emergency services (e.g. centralised control rooms), transport (e.g. intelligent transport systems), energy (e.g. smart meters), waste (e.g. compactor bins and dynamic routing), environment (e.g. flood management), buildings (e.g. building management systems), homes (e.g. controlled smart appliances) and civic (e.g. various apps).

Implementing a "smart city system" cannot avoid collecting urban big data. The urban big data come from three sources: directed (surveillance and public records), automated (digital devices, sensors, etc.) and volunteered (social media, crowd-sourcing, etc.). Different data sources can achieve a diversified range of public and private generation of fine-scale data about citizens and places in rea-time. The data

Fig. 5 Promise of smart urbanism/cities

deluge from transport providers, mobile phone operators, app developers, social media sites, home appliances, remote sensing, emergency services, etc., can then be combined, analysed and acted upon.

The sentient city shall be of automated management with systems to make and enact decisions. It shall employ machine learning to enable some level of artificial intelligence. From a positive aspect, smart cities can render smart economy (innovation, productivity), smart people (more informed), smart living (quality of life), smart environment (green energy, sustainability), smart government (transparency, better service delivery), and smart mobility (intelligent transport systems, efficiency). However, the challenge of smart cities is that they can create the problem of ethical and security issues. The concerns are data privacy, data security, ownership and control of data markets, control creep and hack (Fig. 5).

To alleviate the perils, procedural and normative actions can be taken. Procedural action locate source of the problem in individuals and technical systems. Normative action acknowledges structural power and works towards redistribution and reconfiguration.

Professor Kitchin suggested two solutions. The first solution is some practical tactics, like imposing industry standards and self-regulation and using ethics as a competitive advantage: imposing strong end-to-end encryption and security controls for technology. Providing education and training and setting up "smart city governance, risk and compliance board" are all practical tactics to combat against perils of smart cities.

The second solution is forming a holistic strategy by considering ethics of care, citizenship, social justice, principles, values and ethos. Among them, social justice is a vital concept. Social justice concerns the expected and acceptable ways people are treated and the conditions they live. The theories fall into four types: distribution

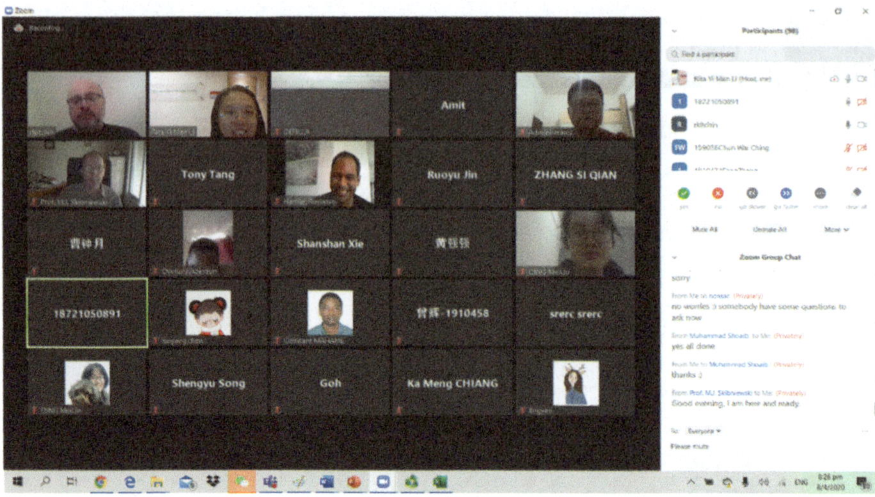

Fig. 6 Participants that joined Prof Rob's discussion section

(fair share), procedural (fair treatment), and retributive (fair punishment for wrongs), restorative (righting of wrongs).

There were 98 participants in the discussion (Fig. 6), and the recording can be found in:

https://www.youtube.com/watch?v=o31mLz0joBo.

His second part of the talk was about publishing (Fig. 7) which can be accessed in https://www.youtube.com/watch?v=OcuIDCGOiBo.

2.4 Miroslaw Skibniewski

Professor Miroslaw J. Skibniewski is the present Editor-in-Chief of the top, international SSCI indexed research journal 'Automation in Construction', which is published by Elsevier. He is A. James Clark Chair Professor at the University of Maryland and previously served as an Assistant Executive Vice President for Academic Affairs and Associate Provost of Purdue University He is on the board of 10 other scientific journals as an editorial member. These scientific journals concern the areas of computer-aided design, automated inspection, facilities management, intelligent control systems etc. He has developed the first and the only one construction automation program and collaborated with major government and industrial organisations in the US and overseas.

Professor Skibniewski gave a speech on "Effective Methods of Publishing Papers in SCI Journals" (Fig. 8), which is essential to researching and publishing artificial

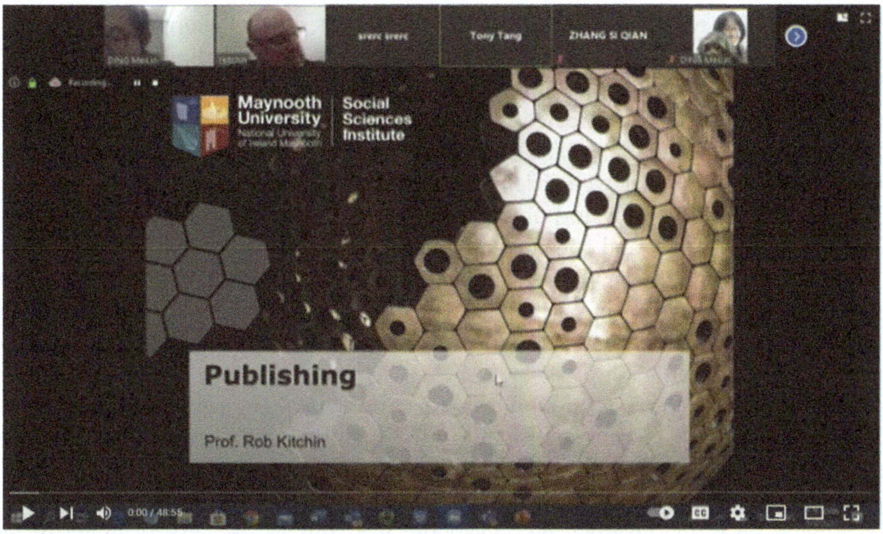

Fig. 7 Prof Rob Kitchin's talk about publishing

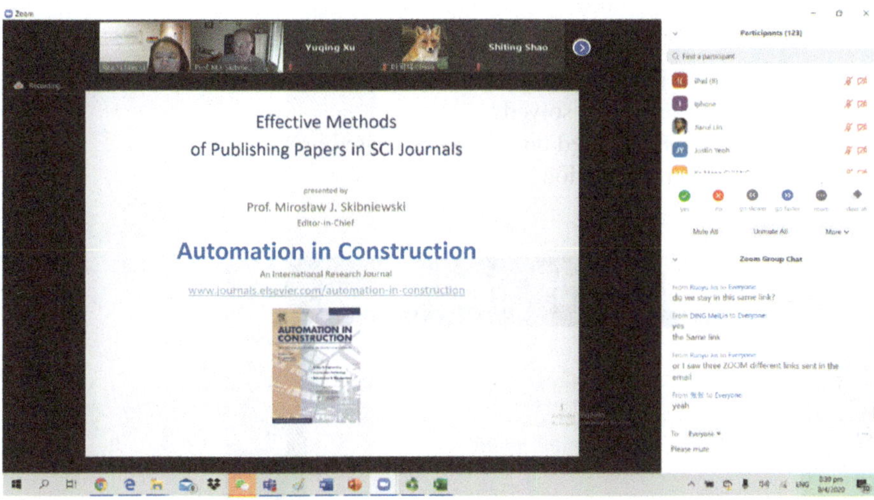

Fig. 8 Professor Skibniewski gave a speech on "Effective Methods of Publishing Papers in SCI Journals"

intelligence in Automation in Construction.

Professor Skibniewski stressed that scientific publications have four roles: registration, certification, dissemination, and preservation. Registration means the timestamp to note who submitted scientific results first officially. Certification is defined as performing peer-review to ensure the validity and integrity of submission. Dissemination refers to providing a medium for discoveries and findings to be shared. Preservation is preserving the minutes and records of science for posterity.

Scientific, technical and medical communities worldwide are united through STM publishing. Today, there are 2,000 STM publishers, 20,000 peer-reviewed journals and 1.4 million peer-reviewed articles, and the trend is increasing exponentially.

For researchers, "ready to publish" means research composing new and original results, plus ready and considered content. Besides full articles, letters or short communications and review papers are also types of manuscripts. Researchers shall also follow the "good listing principle" guideline, which means the corresponding author or the first author, and avoid the "poor listing principle", which means ghost authorship or gift authorship.

Going to titles, Professor Skibniewski gave practical suggestions on "effective manuscript titles". The title shall identify the main issue, adequately describe the content, use the fewest possible words, and not use rarely used abbreviations to attract readers.

Moving on to the abstract, he considered it as this as the advertisement of an article. The abstract shall be attractive, understandable, accurate and specific. A clear abstract will strongly influence whether or not one's work is considered, thus keeping it as brief as possible.

General reasoning in considering the area of research include (Fig. 9):

- Why is this topic important, and for what specific constituents?
- What is the problem to be solved?
- Who has previously worked on similar or related problems?
- What is the research question?

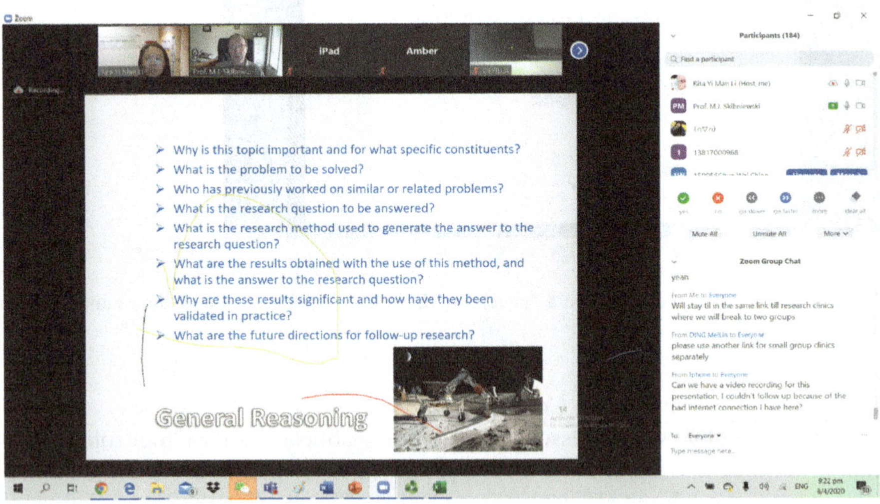

Fig. 9 General reasoning in area of research

- What is the research method used to generate the answer to the research question?
- What are the results significant, and how have they been validated in practice?
- What are the future directions for follow-up research?

Researchers researching artificial intelligence and u-cities shall consider the above issue. One hundred eighty-four participants attend the talk. We can listen to his talk in https://www.youtube.com/watch?v=o31mLz0joBo&t=62s.

2.5 *Simon Beecham*

Prof Simon Beecham joined UniSA in 2006 and was acting Deputy Vice-Chancellor: Research and Innovation from 2019–2020. From 2015 to 2018, Simon was Pro-Vice-Chancellor of the Division of Information Technology, Engineering and the Environment. Before that, he was Head of the School of Natural and Built Environments (2010–2014) and Director of UniSA's Centre for Water Management and Reuse (2006–2010). Simon is a Fellow of Engineers Australia and a Fellow of the Australian Institute of Company Directors.

The outline of "Use of Sensors in Green Infrastructure" presented by Professor Simon Beecham is sensors and smart cities; application of sensors to green infrastructure: permeable pavements, rainwater gardens, living walls; and satellites for remote sensing of urban vegetation (Fig. 10).

Professor Beecham first explained a sensor system, including remote sensing, surveillance system, and dedicated sensor platform. The sensor system can help obtain data, securely and efficiently transport, post-process, and make decisions.

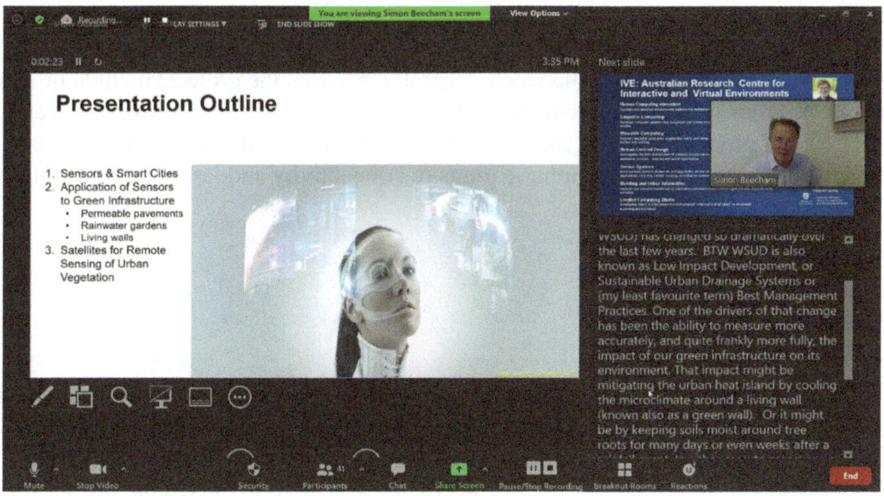

Fig. 10 Presentation outline of Prof Simon Beecham's talk

To use sensors in smart cities, applying computational visualisation to support contemporary design, building and planning of Building Information Modelling (BIM) and City Information Modelling (CIM); computational design analysis; simulation, urban analytics and smart cities; digital humanities and heritage; and design collaboration and behavioural studies.

Professor Beecham continued his presentation by giving examples of how green features can help smart cities. For instance, estimation of security of supply using continuous simulation analysis, estimation of flood control using switch continuous simulation model, use of the embedded sensor in permeable interlocking concrete pavements, sensing soil moisture and groundwater level.

DesignPave software is one of the tools that are utilised in green infrastructure. Geofabric is applied to permeable pavements and coupled with moisture sensors, monitoring boxes, and other rainwater gardens tools for flood control. Estimation of security of supply using continuous simulation analysis and estimation of flood control using switch continuous simulation model were carried out.

Another tool uses embedded sensors in permeable interlocking concrete pavements to sense soil moisture and groundwater level to create a living wall. The sensing equipment included thermocouples (to detect surface temperature), iButtons (to detect air temperature and relative humidity), weather station (to detect air temperature, sunshine and rainfall) and infrared camera model. Microclimate was being sensed around a living wall. The data were analysed for evapotranspiration, water demand and water footprint of urban green spaces.

The methods to estimate evapotranspiration (ET) of urban green spaces include remote sensing-based, ground-based and observational-based.

Professor Beecham continued his presentation by demonstrating two examples: Adelaide parklands and the Aquawatch Australia mission. In Adelaide Park, sensors were used to analyse the intra-annual variation of the greenness of urban vegetation. Three kinds of remote sensing of different resolutions are being utilised, including EOS-1, Aqua (MODIS sensor), Landsat (ETM + and OLI sensor), and worldview 2 (spatial resolution). In the Aquawatch Australia mission, the projects include in-situ sensor monitoring, hydrological network mapping, complex data fusion for water-related operational decisions etc. Next-generation ground stations shall be considered to achieve advanced communications & connectivity and laser communication (Fig. 11).

2.6 George Boustras

Professor George Boustras is the Editor-in-Chief of Safety Science. He has published over 50 Journal and Conference papers and has Edited 3 Books. He is the Editor of the Book Series "The Interface of Safety and Security". He has been appointed by the President of the Republic of Cyprus as Vice President in the Council of Energy Strategy. He sits at the Management Committee of Secure Societies—Protecting

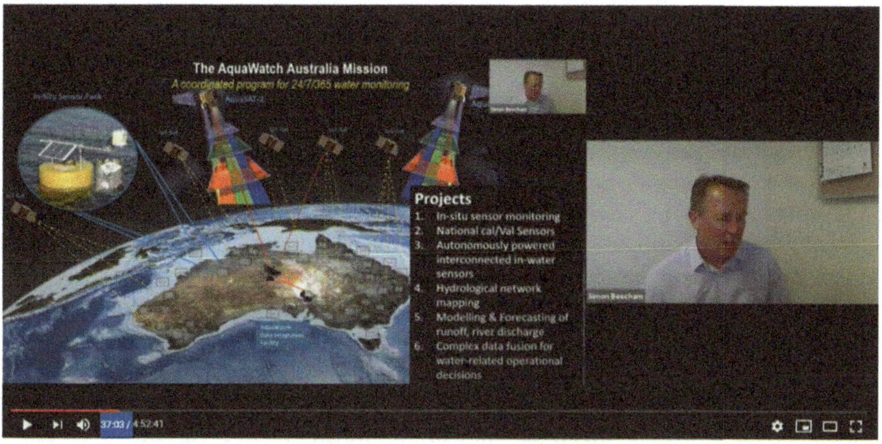

Fig. 11 The AquaWatch Australia Mission

Freedom and Security of Europe, and he is a Member of the Academic Committee of the European Research Network of Critical Infrastructure Protection.

Professor Boustras presented "Emerging Risks in Safety Management" (Fig. 12). In safety management, a resilience assessment tool (RAT) is used. RAT includes web-based user accounts, auto-completion of values using a questionnaire, assessments creation for scenarios' sandboxing, creation and comparison of scenarios, and radar chart as the visual output of the results.

First Responder App is a valuable tool for safety management. A social application allows users to report emergency events; the central operator can then inform first

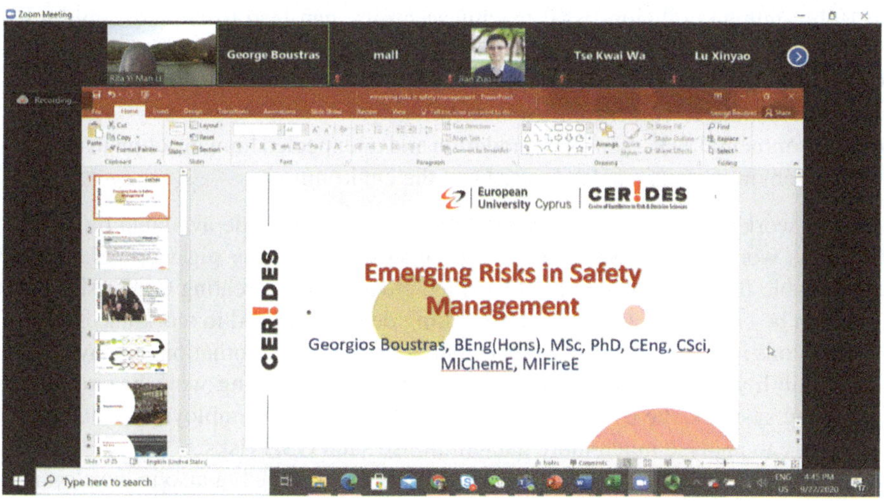

Fig. 12 Emerging risks in safety management

responders near the event. The first respondent that accepts the call can be directed to the event using Google Maps. Medical events can include medical information for the patient. Central operator can then decide to inform the fire department, police department or ambulance.

Professor Boustras mentioned challenges in occupational health and safety (OHS). Work and workplaces are constantly changing with the introduction of new technologies, substances and processes, with changes in the workforce structure and the labour market, as well as with new forms of employment and work organisation. This can create new risks and challenges for the safety and health of employees. There must be expected and addressed in order to ensure safe and healthy workplaces in the future.

He suggested new employment types and new risks would happen. “Crowdsourcing” is a term used for new forms of work in confusion: cooperative economy, “human cloud”, job demand on-demand, digital work. In new forms of work, “remuneration” is defined as paid work that is organised through online exchanges of external assignments of tasks.

Working with computers can cause stress and physical disorder such as eye strain or musculoskeletal problems. There will be severe job insecurity due to which many of the employees employed through payroll do not know if they will have a job the next day. Workers do not communicate directly with the end customer, lack of individual or collective representatives. At the same time, robots allow people to be removed from dangerous physical work and environments with chemical and ergonomic hazards, thus reducing OHS risks for workers.

Although there are significant possibilities for workplace progress and growth in productivity, there are also important OSH safety and health-related questions arising as AI is integrated into workplaces. There are three types of OSH risks in human–robot-environment interactions:

- Robot-human collision risks: machine learning can lead to unpredictable robot behaviour.
- Security risks: robots’ internet links can affect the integrity of software programming, leading to vulnerabilities in security.
- Environmental risks: sensor degradation and unexpected human action in unstructured environments can lead to risks to the environment.

“Gig work” is obtained by using online applications made available by companies. The work can be performed online or offline. Delivery gig workers are held accountable for their speed in an intensified environment, creating OSH risks. Such risks can be caused by wearable self-tracking devices utilised in workplaces.

Solutions include fewer musculoskeletal issues due to automation of heavy work, the possibility to adjust the working time ad hoc, monitoring work hazards using electronic systems, the use of sensors in the workplace, the employee’s clothes, and the ability to analyse much more data minimise such OSH risks.

Safety Management is helpful not only in the workplace but also in developing a smart city (Fig. 13).

Fig. 13 Poster of the talk delivered by Prof Simon Beecham and George Boutras

2.7 *Simon Fong*

Professor Simon Fong graduated from La Trobe University, Australia, with a 1st Class Honours BEng. Computer Systems degree and a PhD. Computer Science degree in 1993 and 1998 respectively. Simon is now working as an Associate Professor at the Computer and Information Science Department of the University of Macau. He is a co-founder of the Data Analytics and Collaborative Computing Research Group in the Faculty of Science and Technology. Before his academic career, Simon took up various managerial and technical posts, such as systems engineer, IT consultant and e-commerce director in Australia and Asia. Dr Fong has published over 432 international conference and peer-reviewed journal papers, mainly in the areas of data mining, data stream mining, big data analytics, meta-heuristics optimisation algorithms, and their applications. He serves on the Journal of Network and Computer

Applications of Elsevier (IF 3.5), IEEE IT Professional Magazine (IF 1.661) and various special issues of SCIE-indexed journals. Simon is also an active researcher with leading positions such as Vice-chair of IEEE Computational Intelligence Society (CIS) Task Force on "Business Intelligence & Knowledge Management", and Vice-director of International Consortium for Optimisation and Modelling in Science and Industry (iCOMSI).

Simon Fong delivered a presentation about "Data Stream Mining for fast Data Analytics and Collaborative Computing Laboratory". In this seminar, Dr Fong has presented the idea of "data streaming and optimisation", "artificial intelligence for COVID-19 breakout", "data streaming mining for fast AI", "IoT platform with data mining capabilities", "smart home", "data collection", "ultimate level of AI", three levels of learning which means "artificial learning, machine learning and deep learning".

Professor Fong considers "data stream mining" and "optimisation" to be meaningful artificial intelligence that is fast, accurate, lightweight and can save lives.

IoT platform is evolving. Start from original "generation 1 automation and control", evolving to "generation 1.5 automation and control + feedback", then add feature and become "generation 1.5 automation and control + feedback (intelligence) + advice". Then the feedback mechanism is replaced by holistic data analysis in "generation 1.5 automation and control + holistic data analytics + advice". It is the era of generation 2, "automation and control + health advice", like having a virtual doctor at home. He also mentioned that the Internet of Things now can be used for our Smart home (Fig. 14).

Professor Fong has given exciting examples to demonstrate how AI analyse data after data collection. It explains how an algorithm works. The concept of algorithm self is "to the ability to think, reason, and learn by itself".

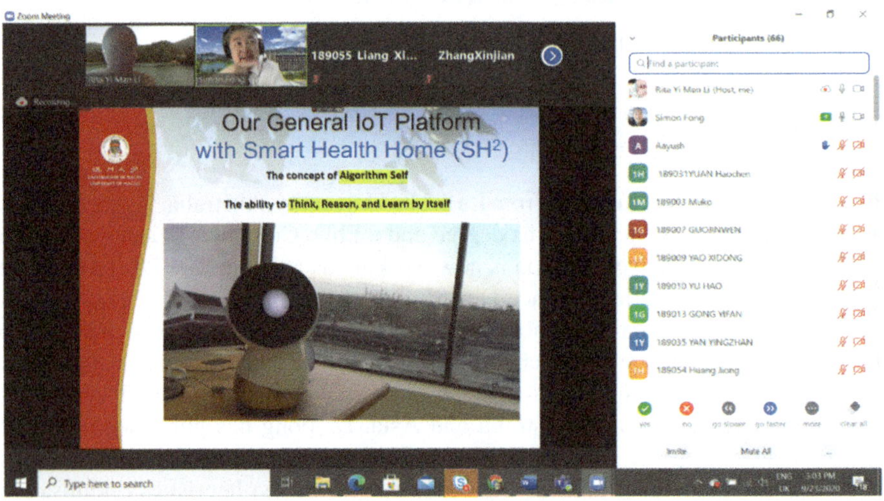

Fig. 14 IoT used for Smart Home

Among all the topics, the most exciting part is his differentiation between "artificial intelligence AI" and "intelligence agent IA". In his opinion, AI is more intelligent than IA. AI is a concept, like a brain, which can think, reason and learn by itself; without this ability, it is merely an IA, not yet AI.

Professor Fong also presented his idea of learning steps, from artificial learning to machine learning, then to deep learning. Artificial learning represents simulated intelligence in machines. It is a subset of data science, and it aims to build machines capable of thinking like humans. Machine learning is the practice of getting machines to make decisions without being programmed. It is a subset of artificial intelligence & data science with its aim to make machines learn through data to solve problems. Deep learning is the process of using artificial intelligence networks to solve complex problems. It is a subset of machine learning, artificial intelligence and data science. It aims to build neural networks that automatically discover patterns for feature detection.

Given the strong demand for accurate and fast AI, FOG computing is one of the tools that add a hierarchy of elements between the cloud and endpoint devices, devices and gateways, to meet the challenges of real-time applications. Human activity recognition is another tool that uses sensors and machine learning algorithms to increase accuracy. Another method is "a scalable data stream mining methodology: stream-based holistic analytics and reasoning in parallel", which aims to design and develop a holistic data-stream-mining system, meaning it can facilitate decision-support of the highest possible accuracy in data-mining data streams.

Prof James Fong's talk can be found at https://www.youtube.com/watch?v=xRud_kvQM4Q.

2.8 *Professor Tan Yigitcanlar*

Tan Yigitcanlar is an eminent Australian researcher with international recognition and impact in urban studies and planning. He is a Professor of Urban Studies and Planning at the School of Built Environment, Queensland University of Technology, Brisbane, Australia. Along with this post, he carries out an Honorary Professor role at the School of Technology, Federal University of Santa Catarina, Florianopolis, Brazil, and the Founding Director position of the Australia-Brazil Smart City Research and Practice Network.

He is an eminent Australian scholar with international recognition, reputation and impact on policy, practice and society. He has been responsible for research, teaching, training and capacity building programs on the fields of urban studies and planning in esteemed Australian, Brazilian, Finnish, Japanese, and Turkish universities. His research addresses contemporary urban planning and development challenges—economic, societal, spatial, governance or technology related in nature. The main foci of his research interests, within the broad field of urban studies and planning, are clustered around the following three interdisciplinary themes.

The first research theme focuses on 'Smart Technologies, Communities, Cities and Urbanism'. In this stream of research, he examines the disruptive effects, along with the beneficial impacts of urban technologies and digital transformation of urban services and infrastructures on our cities and societies. His research in this theme covers, but is not limited to, the following topics: smart urban technology, smart city, intelligent community, innovative precinct, smart urbanism, smart mobility (e.g., autonomous driving, mobility-as-a-service), citizen science (e.g., crowdsourcing, volunteered geographic information), artificial intelligence; and gamification and gamified applications.

Profession Yigitcanlar presented "The Rise of Artificial Intelligence and the Smart and Sustainable City". He gave many examples of how smart cities would be, especially taking Australia as a demonstration.

Smart city is about technology and related to community and policy. Professor Yigitcanlar defined the smart city as "an urban locality that employs digital data and technology to create efficiencies for boosting economic development, enhancing the quality of life and improving the sustainability of the city". He further defined "smart and sustainable city" as "an urban locality functioning as a robust system of systems with sustainable practices, supported by technology, community and policy, to generate desired outcomes and futures for all humans and non-humans". Australia has adopted the concept of "smart city" in several cities like Atlantic.

Professor Yigitcanlar also defined AI as "a software technology with at least one of the following capabilities: perception, decision-making, prediction, automatic knowledge extraction of pattern recognition, interacting communication, or logical reasoning" (Fig. 15).

What is urban AI? Tan considered "UAI is where AIs are embodied in urban technologies, which together are turning cities into autonomous entities operating in an unsupervised manner. Professor Yigitcanlar illustrated how AI could contribute to

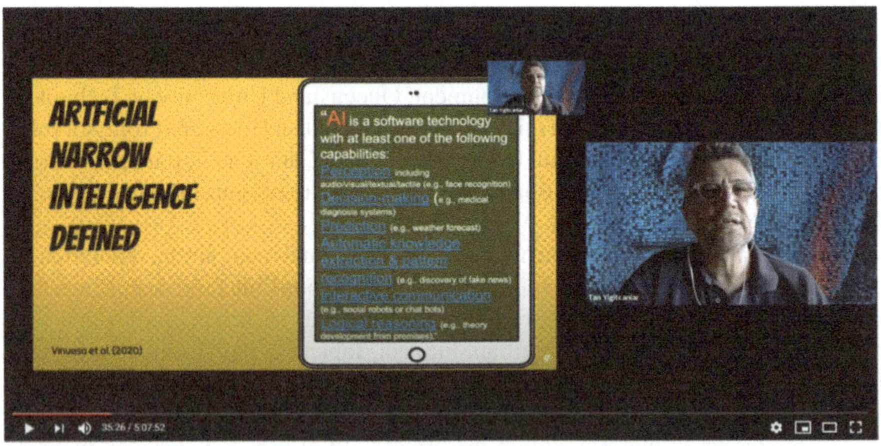

Fig. 15 AI's capabilities as a software technology

Table 2 AI's contribution on economic, social, environmental and government dimensions

AI's contribution in:	Examples
Economic dimension	• It enhances productivity and innovation by automating data management and analysis • Increasing resources and reducing costs through pattern recognition • It supports decision-making by analysing large volumes of data from multiple sources • Drawing conclusions for informed decisions based on logic, reason and intuition via deep learning
Society dimension	• Improving health monitoring via intelligent sensors and analytics • Enhancing health diagnosis outcomes through medical imaging analytics • Providing autonomous tutoring systems to teach algebra and grammar • Offering personalised learning to manage how they progress through learning activities
Environment dimension	• Operationalising smart transport systems via MaaS Optimising energy production and consumption via domestics • Monitoring changes in the natural and built environment via remote sensing with autonomous drones • Predicting the risks of climate change via machine learning algorithms to combine climate models
Government dimension	• Assisting citizen scientists with new technology for informed decisions • Aiding the disaster and pandemic management planning and operations via predictive analytics • Enhancing the operability of surveillance systems via smart places with AIoT • Improving cybersecurity by analysing cyber incident data and identifying potential threats

smart cities from several dimensions, including economy, society, environment, and government (Table 2). He also suggested that automation decision making consisted of 6 different levels (Fig. 16).

Prof Tan Yigtcanlar concludes the whole speech by stating the symbiotic relationship between AI and smart cities. Local modernisation needs modernisation to take full advantage, and things can go wrong if we let algorithms solely make decisions for the government (Fig. 17). The talk has attracted many professors, including Wendell Brase, Associate Chancellor of University of California, Irvine, Professor Patrick Zou from Swinburne and Professor Jian Zou from the University of Adelaide on top of local faculty and students (Fig. 18).

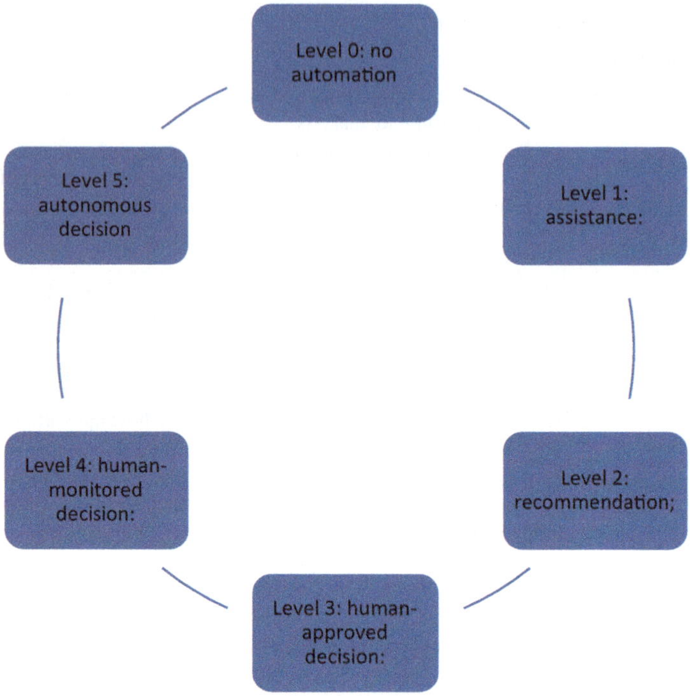

Fig. 16 Different decision automation levels

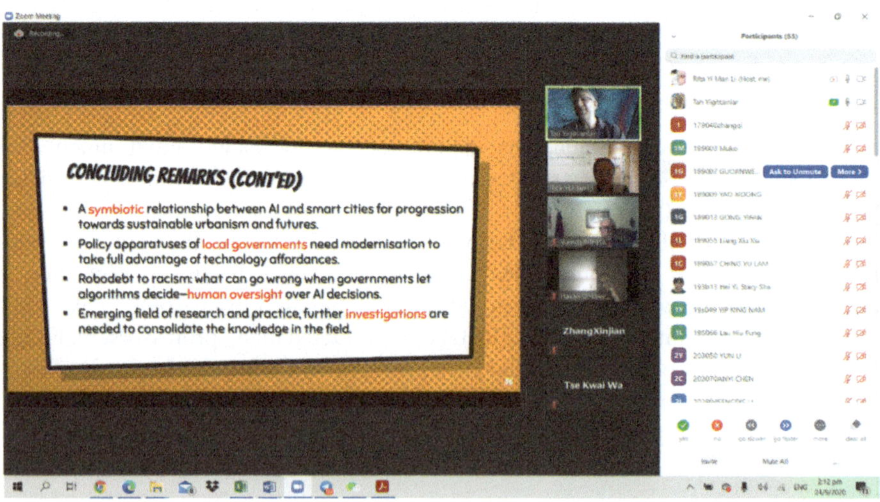

Fig. 17 Artificial intelligence, smart cities and government

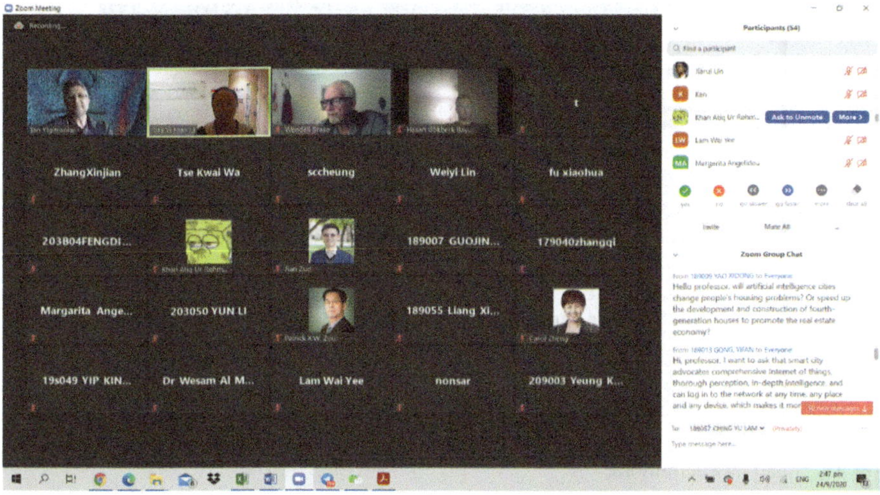

Fig. 18 Participants who joined Prof. Tan Yigtcanlar's seminar

2.9 Zuo Jian

Dr. Jian Zuo is a Professor of Sustainable Construction and Associate Head (International) in the School of Architecture and Built Environment, The University of Adelaide. He is the chief investigator of several projects commissioned and funded by governments in China and Australia, such as the Australian Research Council and Australian Housing and Urban Research Institute. His main research interest relates to achieving good health, safety, and sustainability at both building and urban levels, including using alternative construction methods (e.g., prefabrication), technologies (Building Information Modelling), and policy interventions to achieve better performance environmental and economic sustainability of development projects. His other main research interests include construction and demolition waste recycling, the health and wellbeing of construction workers, smart construction, productivity of building users, indoor environmental quality in buildings, and a mechanism for the generation and mitigation of dust across the lifecycle of buildings.

He has published over 100 peer-reviewed papers in international journals and three ESI Highly Cited Papers. His research is well recognised by the academic society, drawing more than 8000 citations since 2011 with an H-index of 49. He serves as co-editors-in-chief of the Journal of Green Building, Associate Editor of Journal of Cleaner Production, and an editorial board member of Renewable and Sustainable Energy Reviews, International Journal of Construction Management, etc. Professor Zuo enjoys a close relationship with the industry, having been engaged by both Australian and international professional bodies to review green building and low carbon city development tools and resource efficiency issues.

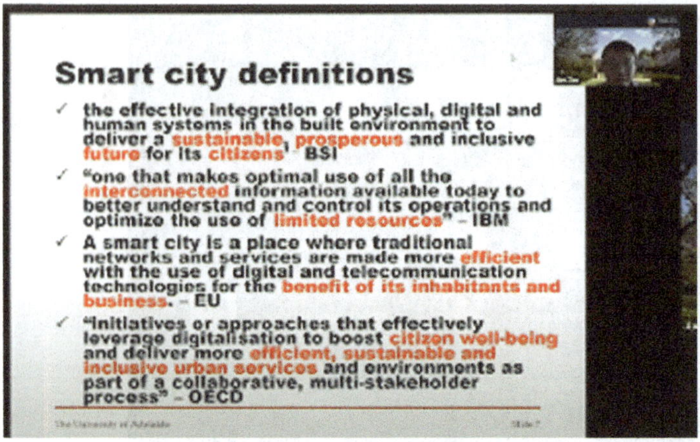

Fig. 19 Smart city definitions

Professor Zuo gave a presentation on "Carbon Neutral City and Smart City go Hand-in-Hand: A Case Study of Adelaide". He mentioned the definitions of smart cities (Fig. 19) and commented that smart cities could assist the following areas:

1. Climate change
2. Sustainability
3. Responsiveness
4. Quality of life
5. Resilience
6. Disaster prevention and recovery: flooding, bushfire

Professor Zuo informed the audience of the six smart cities policies priorities:

1. Jobs and skills
2. Infrastructure and investment
3. Livability and sustainability
4. Innovation and digital opportunities
5. Governance, planning and regulation
6. Housing

He also mentioned the phases of smart cities 1.0 versus 2.0 versus 3.0. In smart cities 1.0, this phase, characterised by technology, encourages the adoption of solutions of cities that are not equipped to understand the implications of the technology or its impact on quality of life. Smart cities 2.0: cities drive the smart city agenda, and technology solutions are seen as enablers to improve quality of life. Smart cities 3.0: the most recent phase of the smart city evolution is grounded in citizen cooperation models which focus on addressing local issues and unlocking local opportunities.

It has become urgent to achieve carbon neutrality due to rising global climate change concerns concerning carbon neutrality. Commitments to the reduction of carbon emissions will be required from all industries. And why do we focus on carbon–neutral cities in particular? Cities account for more than 70% of global co-emissions today. Buildings and construction are responsible for 39% of carbon emissions today. Furthermore, how to achieve carbon neutrality? Approaches include:

1. Adopt a zero-emissions standard for new buildings
2. Build a ubiquitous electric vehicle charging infrastructure
3. Mandate the recovery of organic material
4. Electricity buildings heating and cooling systems
5. Designate car-free and low-emissions vehicle zones
6. Empower local procedures and buyers of renewable electricity
7. Set a city climate budget to drive decarbonisation

He suggested the pathways to carbon neutrality as follows and offered the University of Adelaide as a case study (Fig. 20):

1. Energy-efficient built form: drive investment in the energy efficiency of city buildings and other infrastructure, and green the streetscapes and open spaces;
2. Zero emissions transport: decarbonise transport and change the way we travel
3. Towards 100% renewable energy: continue to drive investment in large-scale renewable energy generation and strategy in the state, and encourage the update of small-scale renewable energy supply and storage in the city
4. Reduce emissions from waste and water: build in South Australia's leading resource recovery and diversion of waste from landfill
5. Offset carbon emissions: reduce emissions rapidly in every way we can, and offset the remaining emissions when appropriate timing has been confirmed with partners

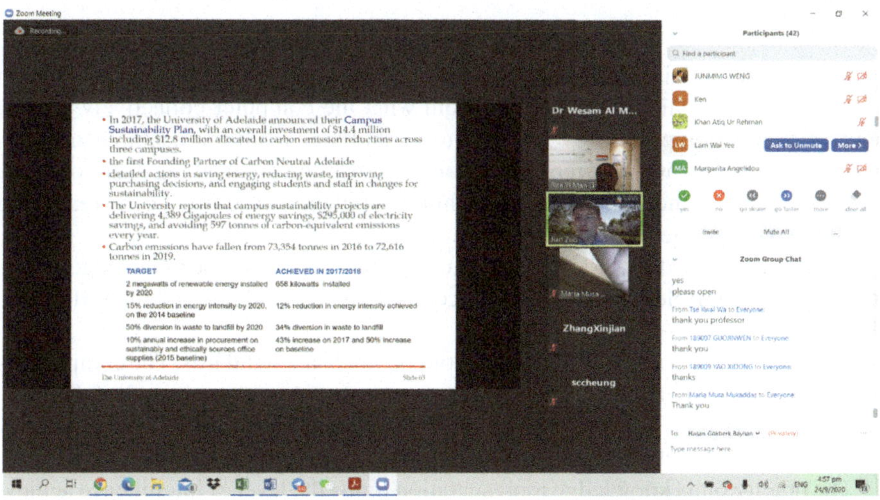

Fig. 20 University of Adelaide plan

2.10 Mark Fox

Dr Fox is a Distinguished Professor of Urban Systems Engineering and a Professor of Industrial Engineering and Computer Science, where his current research applies Artificial Intelligence to Smart Cities. He is the Associate Director (Research) in the School of Cities. He received his BSc in Computer Science from the University of Toronto in 1975 and his PhD in Computer Science from Carnegie Mellon University in 1983.

From 1981 through 1987, he founded and led the Robotics Institute's Intelligent Systems Laboratory, and from 1987 through 1991, he co-founded and led the Center for Integrated Manufacturing Decision Systems. In 1993 Dr Fox co-founded Novator Systems Ltd., a pioneer in out-sourced e-Retail services and software. In 1984 he co-founded Carnegie Group Inc., one of the first companies to apply Artificial Intelligence to solving engineering, manufacturing, and telecommunications problems. He is a Fellow of the Association for the Advancement of Artificial Intelligence (AAAI), the Institute of Electrical and Electronics Engineers (IEEE), and Canada's Engineering Institute.

Mark Fox presented "An Introduction to Ontologies" He considered Ontology design a way of thinking. The ontology engineering process includes:

1. Determine domain and scope
2. Determine the competency questions
3. Consider reusing existing ontologies
4. Enumerate important terms
5. Identify classes and structures as a taxonomy
6. Define classes using properties
7. Define instances
8. Validate using competency questions

Knowledge and its representation are essential for computers daily decision making. It is a sustainable advantage from what the computer collectively knows, how efficiently computer uses what it knows, and how readily it acquires and uses new knowledge. Knowledge representation is a substitute for the thing itself to enable the robot to determine consequences by thinking rather than acting (i.e. by reasoning about the world rather than taking action in it).

He also mentioned SPARQL, similar to SQL but adapted to the knowledge graph. Its endpoint is a triplestore that can be accessed via http and queried with SPARQL (Fig. 21).

His work focuses on transportation ontologies and data standards. The components of ontology are (1) knowledge graph (classes, properties, instances); (2) definitions and constraints (class definitions, automated classification); and (3) micro-theory (axioms/rules, deduction—answering questions) (Fig. 22).

The fidelity of its knowledge representation limits the level of intelligence achievable by software. In principle, perfect fidelity is not possible because anything other than the thing itself is theoretically different from the thing. In general, the

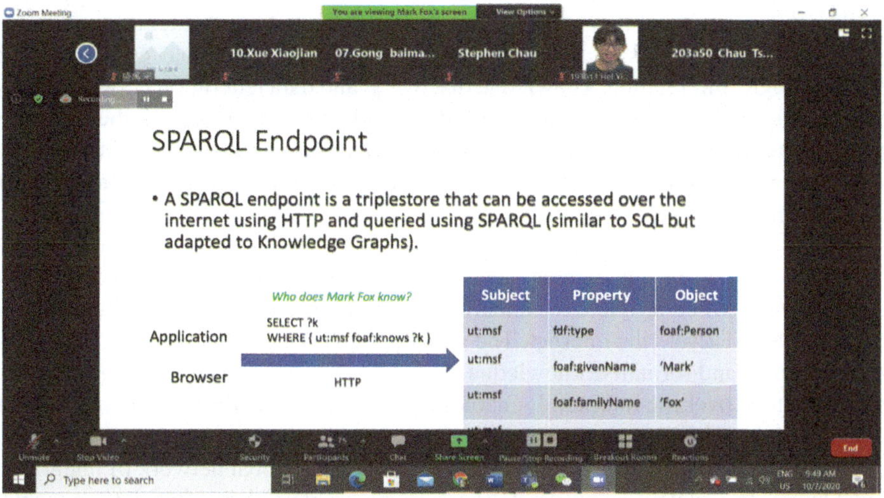

Fig. 21 SPARQL Endpoint

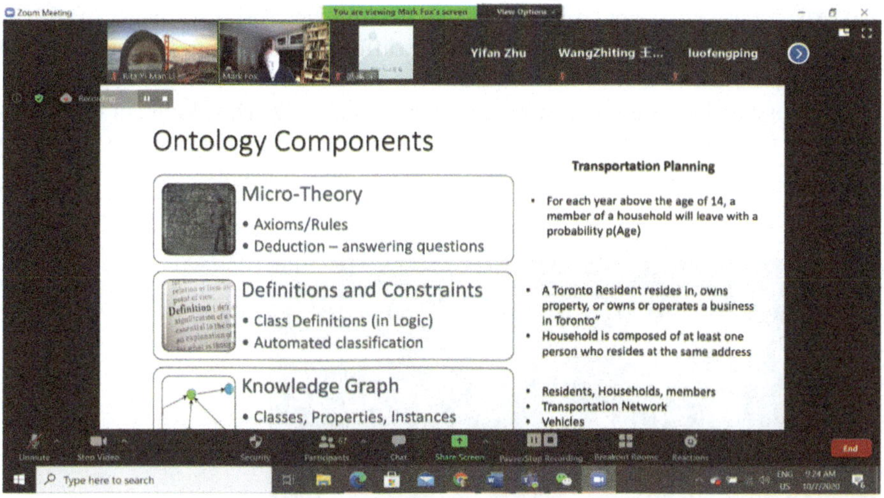

Fig. 22 Ontology components

only completely accurate representation of an object is the object itself. All other representations are inaccurate; they inevitably contain simplifying assumptions.

All reasoning systems, even predictive analytics, such as neural nets, depend upon a knowledge representation "fit for purpose". The question is whether the representation is sufficient for the reasoning to be performed. For example, is it fit for purpose? The risk of imperfect fidelity is that any reasoning has done will eventually lead to incorrect conclusions, independent of the soundness of the reasoning process.

Semantic interoperability is defined as the ability of computer systems to exchange data with unambiguous shared meaning. It is required to enable computational machine logic, inferencing, knowledge discovery, and data federation between information systems. Integrating data across the web poses a challenge as the meaning of the attributes and values differ both within organisations and across the internet. The concern is laid on the packaging of data and the simultaneous transmission of the meaning.

To integrate data from two different locations on the internet, we need to agree on "terms": what are the standard classes and properties. One of the solutions is semantic web which provides a standard for the semantic representation and linking data distributed across the internet. The semantic web is the application of ontologies to standardise and formalise knowledge across the internet. It enables access to data, by computers, over the internet. It enables a shared understanding of the meaning of data wherever it is located and whoever created it.

The tools are now in place, after 70 years, to create a global network of integrated knowledge. The goal is to identify, define, formalise and validate urban data concepts. In particular, the objectives are to (Fig. 23):

1. Identify candidate concepts and their use cases through a detailed analysis of existing urban vocabularies, ontologies and city enterprise software
2. Reduce the candidate concepts from the different sources into a minimal set of core concepts
3. Manage and curate an open process where Standards Development Organizations (SDOs) and other experts can comment on the core concepts (defined in step 2) and post modifications, use cases and new concepts

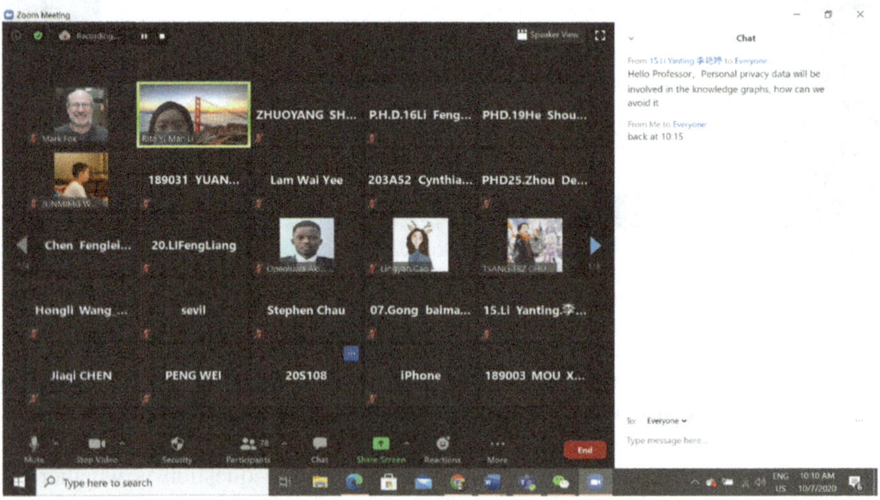

Fig. 23 Participants who joined Mark Fox's talk

4. Formalise and evaluate the resulting concepts from step 3
5. Promulgate the results through the International Standards Organisations.

His talk can be found in https://www.youtube.com/watch?v=3nw9gvrELFo&t=5s.

2.11 John Macintyre

Professor John Macintyre is the Dean of the Faculty of Applied Sciences and Pro-Vice-Chancellor at the University of Sunderland. He has worked at the University of Sunderland since 1992, having graduated from the University with a First-Class Honors Degree in Combined Science (Computer Science and Physiology). He then completed a PhD in applied artificial intelligence, focusing on the use of neural networks in predictive maintenance, which was awarded in 1996.

During the 1990s, John established a research centre—the Centre for Adaptive Systems—at the University, which became recognised by the UK government as a Centre of Excellence for applied research in adaptive computing and artificial intelligence. The Centre undertook many projects working with and for external organisations in the industry, science and academia, and three years ran the Smart Software for Decision Makers programme on behalf of the Department of Trade and Industry. He has successfully supervised PhDs in fields ranging from neural networks, hybrid systems, and bioinformatics through to lean manufacturing, predictive maintenance, and business and maintenance strategies.

Professor Macintyre presented "The Future of AI". AI offers enormous potential to support agile clinical decision making, ensuring patients receive the most appropriate support and leading to better patient outcomes. AI can assist in diagnostics, treatment, prevention measures, and future protection stages. At present, AI experienced exponential growth. Such increase in application and development may be at a tipping point similar to which faced genetic science two decades ago (Fig. 24).

There is some concerns around false expectations and over-selling the ability of AI—and the potential for a backlash. However, there is a real risk that those who don't understand what AI is, how it works, and what it can and cannot do, think AI is the solution to their problems. Those promoting AI applications and solutions need to:

- Build predictive models without sufficient past data also raise risks of unfit models for the purpose.
- Consider ethics is one of the concerns for AI. Privacy concerns and personal data are being discussed.
- Not to over-promise and under-deliver—which could damage confidence in AI in the medium term.

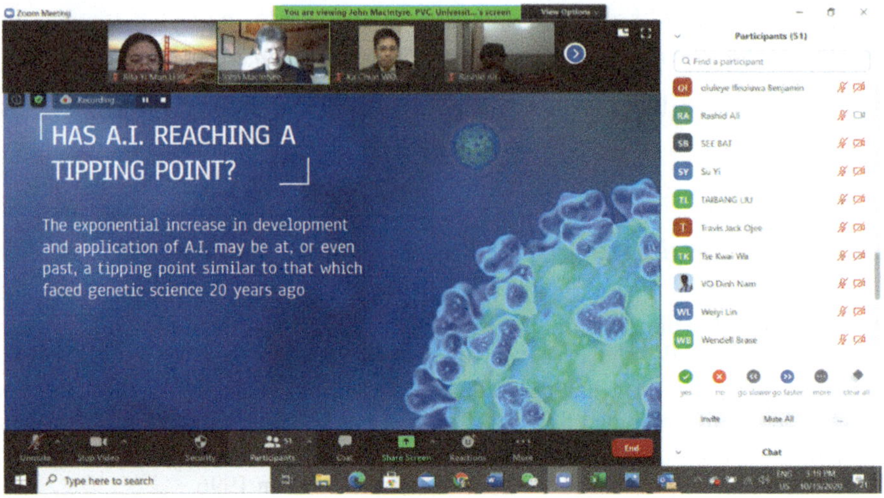

Fig. 24 Has AI reached a tipping point

There are risks of AI. As AI develops in terms of capability, power, and the extent of use across every aspect of human life, the potential for accidental or deliberate misuse increases exponentially. Many examples of AI applications give rise to difficult ethical and moral questions.

Ethical AI is more than "explainable AI". We need to encourage wide-ranging thinking, discussion, debate, and application of ethical design and development of AI. This should be a pragmatic debate leading to specific, useful, practical things AI development teams can use to help them build ethics into their AI. The fact that AI techniques have developed the power and capability to do amazing things does not mean that just because we"can" do something, then we "should" do it.

Meanwhile, much work is being done worldwide on developing guidelines, governance frameworks, regulations, and potentially laws reflecting AI development. The AI community must engage with these developments in an open way to ensure they are meaningful and helpful. We should think of trust, bias, fairness, and accountability to build ethical AI.

His talk can be found in https://www.youtube.com/watch?v=Hy01QZTEJPc&t=523s.

2.12 *Thad Starner*

Thad Starner is a Professor of Computing at Georgia Tech and a Staff Research Scientist in Google's Research and Machine Intelligence division. Thad teaches Artificial Intelligence at Georgia Tech and in Udacity's AI Nanodegree with Udacity founders and AI pioneers Sebastian Thrun and Peter Norvig. Dr Starner is perhaps

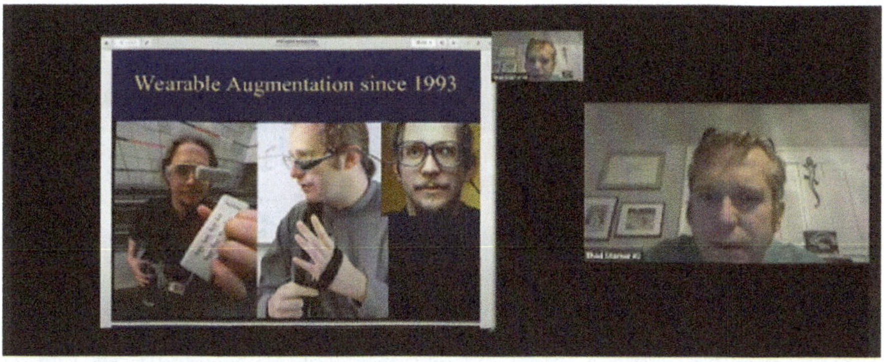

Fig. 25 Wearable augmentation since 1993

most famous for his work on wearable computers, and he was one of the architects of Google Glass, whose third edition was released in 2019.

Thad's other current projects include phrase-level sign language recognition, making wearable computers for communicating with working dogs, creating gloves that teach skills like playing the piano with little to no attention, and helping marine mammalogists learn to communicate with wild dolphins through AI. He also mentioned "Symbiotic AI," which uses wearable devices to teach computers how to live in a human world.

During the presentation, he demonstrated symbiotic AI functions via several videos. Wearable augmentation was one of the oldest features developed since 1993 (Fig. 25). Augmenting conversations used the dual-purpose speech to construct phrases to inform conversational partners and control the computer simultaneously. There were leverage displays and "push-to-talk" to tell the computer when to listen. Speech recognition was only good enough to listen to the user, and hence privacy could be maintained.

Live technology transcribes and translates everything that comes to our phones (Fig. 26). "DialogTabs" can transform audio to the transcript, and the caption helps those with hearing problems & deaf, and the technology of "augmenting reality with crowdsourcing" can help people with visual impairment. Among it, "Openglass" generates visually impaired user test; "Question Answering" allows user to ask the question, receive answer from Twitter/Mechanical Turk; "Memento" let users sighted under associates objects/scenes with annotations and visually impaired user is read annotations in real-time; "Viswiz" makes it possible to scene recognition.

A symbiotic approach to intelligence is interweaving a computer with something already intelligent. Wearable uses machine learning to discover the human world and interact in it. The wearable assistant becomes more intelligent, proactive, and socially graceful in return. For instance, put cameras where eyes are, put microphones where ears are, and put motion sensors at hands (Fig. 26).

Some examples of location/schedule discovery are illustrated during the presentation. It refers to meaning and prediction from unsupervised learning with low-level

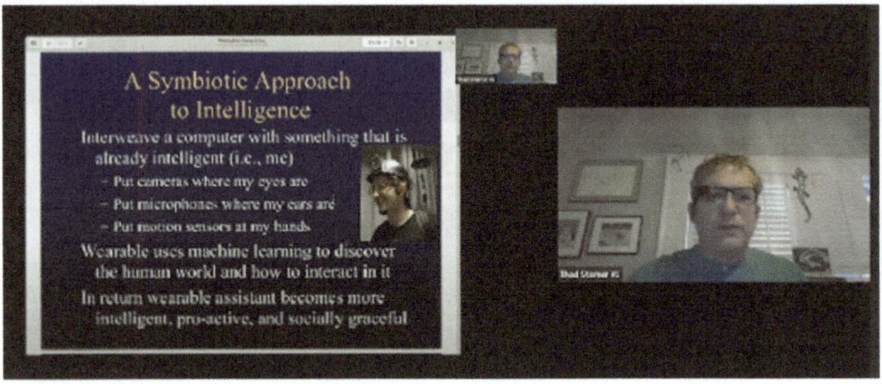

Fig. 26 A symbiotic approach to intelligence

sensing. Meaningful locations are first discovered, and places are clustered into locations, and then locations can further be divided into multiple scales of sublocations. Prediction of user movement can thus be predicted by giving each location a unique ID and replacing each "place" in the original list with ID. The predictions can be classified further as first order or second order.

In an example of "a wrist-worn fitness sensor that discovers new exercise", sensor data collected from thirty-two dumbbell exercise routines were inputted, and the output was the six dumbbell exercises and recognisers. Another example was "discovering signs in American Sign Language", where objects were recognised by how they were manipulated.

True artificial intelligence can only be achieved by machines with sensory and motor skills and are connected to the world through a body. Symbiotic AI uses a "big data" approach, collets first-person data from on-body sensors, processes locally to optimise privacy, network transmission and power. It creates useful and usable intelligent agents leveraging AI that becomes increasingly sophisticating AI that becomes increasingly sophisticate living in the human world.

2.13 Dementri Terzopoulos

Demetri Terzopoulos (PhD'84 MIT), Chancellor's Professor of Computer Science at UCLA, holds the rank of Distinguished Professor and directs the UCLA Computer Graphics & Vision Laboratory. He is or was a Guggenheim Fellow, a Fellow of the ACM, a Fellow of the IEEE, a Fellow of the Royal Society (FRS) of London, a Fellow of the Royal Society of Canada (FRSC), a member of the European Academy of Sciences and the New York Academy of Sciences, and a life member of Sigma Xi.

As a highly cited author in engineering and computer science, his many awards include an Academy Award for Technical Achievement from the Academy of Motion Picture Arts and Sciences for his pioneering research on physics-based computer animation, and the Computer Pioneer Award and inaugural Computer Vision Distinguished Researcher Award from the IEEE for his pioneering and sustained research on deformable models and their applications.

Professor Dementri Terzopoulos gave a presentation on "Biometric Simulation" and demonstrated the future of anthropometric robotics (Fig. 27). There were many demonstrations on how machines have to learn expression from human beings by analysing their muscles and movements. The theme is a human biometric simulation and neuromuscular and sensorimotor control deep learning. The face is a complex biomechanical structure that is difficult to intimate.

In a physics-based face model, one has to consider expression (facial action coding system), control (coordinated facial actuator commands), muscles (contractile muscle fibres exert forces), physics (muscle forces deform synthetic tissue), geometry (expressive facial deformations) and images (rendering by graphics pipeline).

It requires deep learning of neuromuscular control on a comprehensive, realistic biomechanical musculoskeletal model of the human body, including almost all the articular bones and skeletal muscles, and a volumetric finite element soft tissue model.

Multi-physics simulation includes force-coupled, interleaved time-integration across three specialised simulators. Articulated multi-body method is chosen for

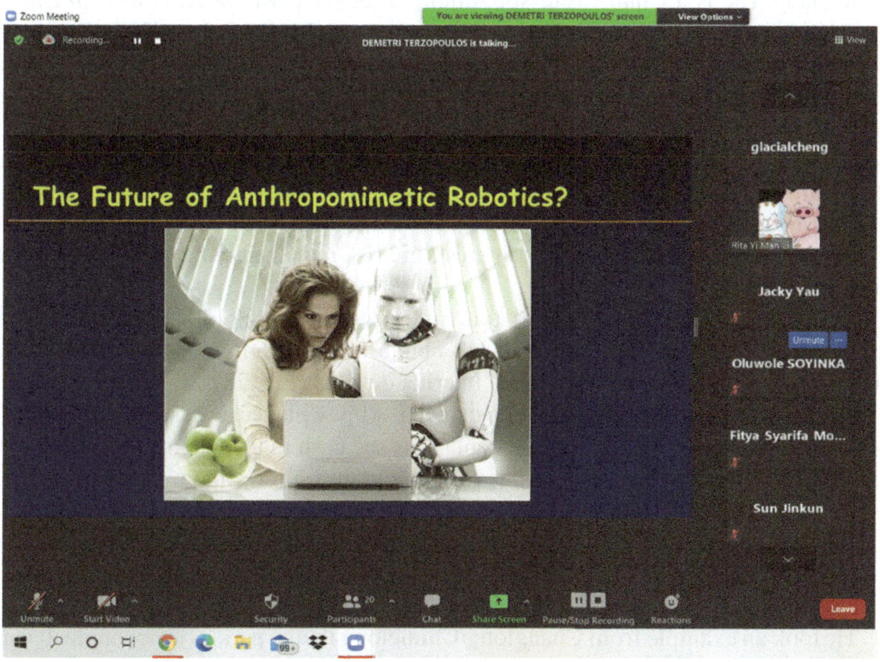

Fig. 27 The future of anthropometric robotics

rigid/articulated body simulation (i.e. bones / skeleton). Lagrangian finite element quasi-incompressible elasticity is utilised for deformable solid body simulation (i.e. flesh), and eulerian fluid simulation with particle level-set method is used for fluid simulation (i.e. water in the human body).

The only viable approach for non-locomotional, task-oriented musculoskeletal control is learning-based neuromuscular control, which is similar to how the human brain does it. It required deep learning of the biochemical human musculoskeletal model synthesises its training data and "trains itself". Moreover, sensorimotor control refers to sensory perception and motor subsystems working together.

Moving to whole-body neuromuscular motor control system, there are six neuromuscular motor controllers and twelve trained deep neural networks for musculoskeletal complexes of core (torso), cervicocephalic, two arms and two legs.

Even for eye movement, it is not simple. A detailed eye model consists of muscles, cornea, iris and lens. For photoreceptor irradiance computation, it includes ray tracing the ocular organs, the cornea refracts light rays, iris and finite-aperture pupil and deformable lens refractors rays. Thus, he considers that there is still a long way for AI development.

2.14 Xiaoguang Yue

Xiaoguang Yue is a Chair Professor/Adjunct Professor/ Research Fellow in China (WBU), Thailand (RMUTR), India (SOA), Cyprus (EUC), Italy (GFU), Romania (RAU), Portugal (IPP) and Egypt (ASU), managing a research team of more than 300 scholars, academicians, and professors. He is the adviser of the Rector of the European University of Cyprus. He is also the Founder and Chairperson of the International Engineering and Technology Institute (IETI), with more than 5000 members. There are five Nobel Prize Laureates and two Turing Award Laureates among them. He has hosted or participated in more than 20 projects and published more than 100 academic papers. His research interests are risk management, sustainability, and intelligent information processing.

Prof Yue mentioned that smart tourism uses new technologies such as cloud computing and the Internet of Things to obtain travel information via the internet, mobile internet and portable terminal internet devices (Fig. 28).

During the pandemic, mobile payments like Alipay played an essential role. First, we frequently use it to avoid people contacting the cash in the banknotes, which can rip. A facial recognition payment via face scanning allows them to recognise us without removing the mask.

Tourists may scan the code when they enter the park or use facial recognition technology. This information will then be uploaded to the database, and then cameras in the gate can automatically match who enter the park.

He took an example from Changlong/Chimelong theme park as an example. It can recommend which place to go according to the rankings of the children or parents. Before we enter the park, we can utilise the online dynamic queueing function to

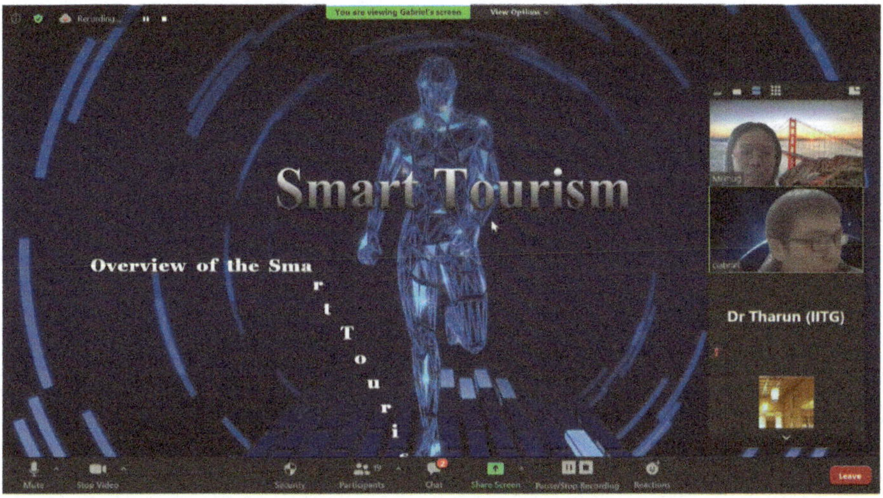

Fig. 28 Professor Yue's presentation of Smart Tourism

check when we need to wait. The smart system will also send the tourists a reminder 15 min before the start of the performance (Fig. 29).

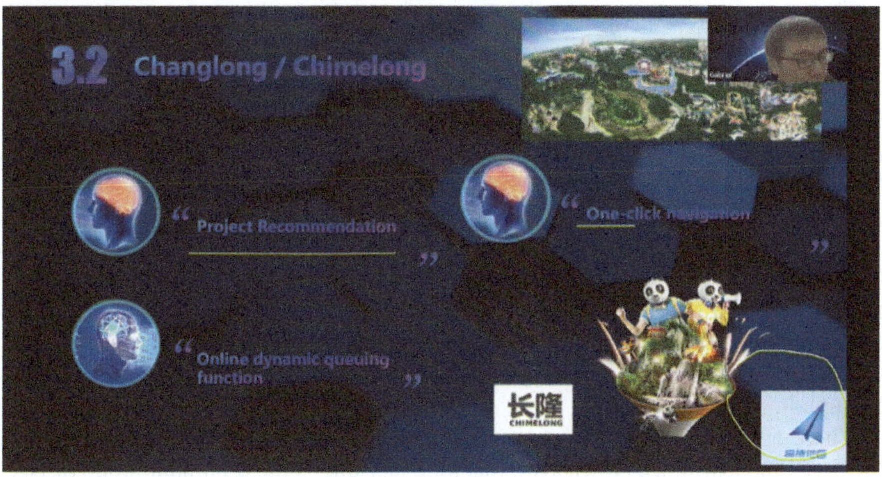

Fig. 29 Changlong/Chimelong smart park's smart system

2.15 Patrick Wang

Prof Patrick S.P. Wang, PhD, is Fellow of IAPR, ISIBM and WASE, IEEE Outstanding Achievement Awardee, Tenured Full Professor, Northeastern University, USA, iCORE (Informatics Circle of Research Excellence), Visiting Professor, University of Calgary, Canada., Otto-Von-Guericke Distinguished Guest Professor, Magdeburg University, Germany, Zijiang Visiting Chair, ECNU, Shanghai, China., NSC Visiting Chair Professor, NTUST, Taipei, Taiwan, Honorary advisory Professor of several key universities in China, including Sichuan University, Xiamen University, East China Normal University, Shanghai, and Guangxi Normal University, Guilin, Prof. Wang received his BSEE from National Chiao Tung University (Jiaotong University), MSEE from National Taiwan University, MSICS from Georgia Institute of Technology, and PhD, Computer Science from Oregon State University.

Published over 26 books, 160 technical papers, 3 USA/European Patents, in PR, AI, TV, Cybernetics and Imaging, Founding Editor-in-Chief of IJPRAI (International Journal of Pattern Recognition and Artificial Intelligence), and Book Series of MPAI, WSP., Co-Chief Editor, IJPRAI and MPAI Book Series, WSP Northeastern University, Boston, MA, USA., In addition to his technical interests, Dr Wang also published a prose book, "Harvard Meditation Melody".

Professor Patrick Wang presented "Intelligent Pattern Recognition, Image Processing and Applications". He defined artificial intelligence as "machine learning of human behaviours to solve a problem that normally can be solved by a human being".

He shared that there are two purposes of AI. One is to use the power of computers to augment human thinking. Robotics and expert systems are major branches of that. The other is to use a computer's artificial intelligence to understand how humans think (in a humanoid way). If one test programs not merely by what they can accomplish, but how they accomplish it, cognitive science is being done. People are using AI to understand the human mind.

Application of AI includes military use as proved in the Persian Gulf War in the early 1990s. Cruise missiles were equipped with AI-related technologies such as Robotics and Machine Vision. Another commonly known application of AI is in chess competitions. In 1997, a computer named "Deep Blue", whose capability was to process 200 million chess positions per second, has defeated the world chess champion.

There are two primary AI approaches, logic and rules-based and pattern-based (i.e. machine learning). Nowadays, machine learning is the dominant mode of AI.

AI learns from data and improves over time. In machine learning, algorithms find patterns in data and infer rules independently. These patterns can be used for automation or prediction. Machine learning steps are learning, pattern detection, data, and self-programming. Examples of pattern-based approaches (machine learning) include optical character recognition, speech recogniser, machine aided assistant, ATM, etc. The challenge of machine learning is that it can only help if good data is available. A typical data handling paradigm is as follows: data, processing information, knowledge, actionable intelligence, decision support and executable actions.

Moving on to Big Data, it is defined as "a broad term for data sets so large or complex that traditional data processing applications are inadequate". Challenges include image analysis, especially biometrics images including audio and video data, capture, data curation, search, sharing, storage, transfer, visualisation, and information privacy. Big data is being generated by everything everywhere at all times. Every digital process and social media exchange produces it. Systems, sensors and mobile devices transmit it. Big data is arriving from multiple sources at an alarming velocity, volume and variety. To extract meaningful value from big data, we need optimal processing power, analytics capabilities and skills.

In the end, Professor Wang shares some areas of future research, such as automatic recognition, biopharma, and establishing an extensive imaging database. He encouraged that research breakthroughs often come about from combining different scientific ideas.

Professor Wang's presentation can be found in https://www.youtube.com/watch?v=b4256D82Fpk.

2.16 *Add on Workshop and summary of all the seminars and workshops*

As many workshop participants do not have strong programming background, we have invited H2o.ai to deliver a workshop which shows the procedure of using this software to conduct AI prediction. Summary of seminars and workshops for this project is tabled as follows:

	Date	Name	Topic
(1)	April 6 2020	Virginia Dignum	Responsible artificial intelligence
(2)	April 6 2020	Michael Batty	Artificial intelligence and smart cities
(3)	April 8 2020	Rob Kitchin	The programmable city
(4)	April 8 2020	Miroslaw Skibniewski	Effective Methods of Publishing Papers in SCI Journals
(5)	September 22 2020	Simon Beecham	Use of Sensors in Green Infrastructure
(6)	September 22 2020	George Boustras	Emerging Risks in Safety Management
(7)	September 23 2020	Simon Fong	Data Stream Mining for fast Data Analytics and Collaborative Computing Laboratory
(8)	September 24 2020	Tan Yigitcanlar	The rise of artificial intelligence and the smart and sustainable city
(9)	September 24 2020	Zuo Jian	Carbon neutral city and smart city go hand-in-hand: a case study of Adelaide
(10)	October 7 2020	Mark Fox	An introduction to ontologies
(11)	October 15 2020	John Macintyre	The Future of AI
(12)	December 18 2020	Thad Starner	Symbiotic AI

(continued)

(continued)

	Date	Name	Topic
(13)	December 18 2020	Jong Hang Siong	H2o.ai workshop
(14)	December 29 2020	Dementri Terzopoulos	Biometric simulation
(15)	December 29 2020	Xiaoguang Yue	Smart Tourism
(16)	December 29 2020	Patrick Wang	Intelligent Pattern Recognition, Image Processing and Applications

3 Conclusion

To conclude, our event series has informed our audience as per the latest development trends, challenges and prospects in AI and u-city, identified various AI and ubiquitous tools that can be used in research projects and reviewed the current state of art applications of AI for u-city.

3.1 The Latest Development Trends, Prospects and Challenges in AI and U-city

Professor Dignum mentioned that AI identifies images, text and video data, extrapolates patterns, and acts based on the pattern. Prof Beecham mentioned that the Aquawatch Australia mission includes sensor monitoring, hydrological network mapping, complex data fusion for water-related operational decisions. Prof George Boutrass recommended the First Responder App as a valuable tool for safety management. It allows users to report emergency events; the central operator then informs the first responders to accept the call and direct them to the event using Google Maps. Sensors are used in the workplace and employees' clothes to monitor work hazards using electronic systems. All these allow us to obtain more data for analysis and minimise occupational safety and health (OSH) risks. Professor Yigitcanlar illustrated how AI could contribute to the ubiquitous city from several dimensions, including economy, society, environment, and government.

Prof. George Boutrass suggested three types of OSH challenges in human–robot-environment interactions: (1) Robot-human collision risks: machine learning can lead to unpredictable robot behaviour. (2) Security risks: robots' internet links can affect the integrity of software programming, leading to vulnerabilities in security. (3) Environmental risks: sensor degradation and unexpected human action in unstructured environments may cause risks to the environment. Prof Mark Fox said the fidelity of its knowledge representation limits the level of intelligence that a piece of software can achieve. Prof John Macintyre suggested that there are risks of AI.

As AI develops, the potential for accidental or deliberate misuse increases exponentially. Increasing examples of AI applications give rise to difficult ethical and moral questions.

3.2 AI and Ubiquitous Tools that Can Be Used in Research Projects

Professor Michael Batty demonstrated the software Quant developed by his team to analyse city data. Professor Simon Beecham mentioned his software DesignPave as a tool for green infrastructure. Geofabric was applied to permeable pavements with moisture sensors, monitoring boxes, and other rainwater garden tools for flood control. Prof Mark Fox mentioned SPARQL, similar to SQL but could be adapted to the knowledge graph. Its endpoint is a triplestore that can be accessed on the internet by using http and queried with SPARQL.

3.3 The Current State of Art Applications of AI for U-city

Dr. Fong suggested that human activity recognition uses sensors and machine learning algorithms to increase accuracy. Prof Thad Starner showcased his wearable machine learning inventions, which applied AI for u-city. "DialogTabs" transformed audio to transcript for the deaf. "Augmenting reality with crowdsourcing" helped people with visual impairment. "Openglass" was developed for the visually impaired user; "Question Answering" allowed a user to ask the question, receive an answer from Twitter/Mechanical Turk; "Memento" let users sighted under associated objects/scenes with annotations and visually impaired users read annotations in real-time. Wearable uses machine learning was used to discover the human world and interact in it. Thus, wearable assistants became more intelligent, proactive, and socially graceful in return. Professor Yue used Changlong/Chimelong theme park where the AI system recommends which place to go.

Acknowledgements This work described in this paper was fully supported by a grant from the Research Grants Council of the Hong Kong Special Administrative Region, China (Project No. UGC/IIDS15/E01/19). IETI Hong Kong, Artificial Intelligence Society of Hong Kong Foro, Hong Kong Computer Society

References

Alanne, K., & Cao, S. (2019). An overview of the concept and technology of ubiquitous energy. *Applied Energy, 238*, 284–302.

Da Costa, C. A., Pasluosta, C. F., Eskofier, B., Da Silva, D. B., & Da Rosa, R. R. (2018). Internet of health things: Toward intelligent vital signs monitoring in hospital wards. *Artificial Intelligence in Medicine, 89*, 61–69.

Ghaemi Rad, T., Sadeghi-Niaraki, A., Abbasi, A., & Choi, S. M. (2018). A methodological framework for assessment of ubiquitous cities using ANP and DEMATEL methods. *Sustainable Cities and Society, 37*, 608–618.

Li, R. Y. M. (2013). The usage of automation system in smart home to provide a sustainable indoor environment: A content analysis in Web 1.0. *International Journal of Smart Home, 7*, 47–60.

Peha, J. (2019). Robots, telework, and the jobs of the future. *Science, 363*, 38.

Serrano, E., & Botia, J. (2013). Validating ambient intelligence based ubiquitous computing systems by means of artificial societies. *Information Sciences, 222*, 3–24.

Wall, L. D. (2018). Some financial regulatory implications of artificial intelligence. *Journal of Economics and Business, 100*, 55–63.

The manufacturer's authorised representative in the EU is Springer Nature Customer Service Centre GmbH, Europaplatz 3, 69115 Heidelberg, Germany. If you have any concerns regarding our products, please contact ProductSafety@springernature.com

Printed and bound by CPI Group (UK) Ltd, Croydon, CR0 4YY
30/09/2026
02225329-0004